INNOVATIVE SITE REMEDIATION TECHNOLOGY

THERMAL DESORPTION

One of an Eight-Volume Series

Edited by
William C. Anderson, P.E., DEE
Executive Director, American Academy of Environmental Engineers

1993

Prepared by WASTECH®, a multiorganization cooperative project managed by the American Academy of Environmental Engineers® with grant assistance from the U.S. Environmental Protection Agency, the U.S. Department of Defense, and the U.S. Department of Energy.

The following organizations participated in the preparation and review of this volume:

Air & Waste Management Association
P.O. Box 2861
Pittsburgh, PA 15230

American Society of Mechanical Engineers
345 East 47th Street
New York, NY 10017

American Academy of Environmental Engineers®
130 Holiday Court, Suite 100
Annapolis, MD 21401

Hazardous Waste Action Coalition
1015 15th Street, N.W., Suite 802
Washington, D.C. 20005

American Institute of Chemical Engineers
345 East 47th Street
New York, NY 10017

Water Environment Federation
601 Wythe Street
Alexandria, VA 22314

Library of Congress Cataloging-in-Publication Data

Innovative site remediation technology / edited by William C. Anderson
p. cm.
Includes bibliographic references.
Contents: 6. Thermal desorption
ISBN 1-883767-06-7 (v. 6)
1. Soil remediation. I. Anderson, William, C., 1943- .
II. American Academy of Environmental Engineers.
TD878.I55 1993
628.5'5--dc20 93-20786
CIP

The material presented in this publication has been prepared in accordance with generally recognized engineering principles and practices and is for general information only. This information should not be used without first securing competent advice with respect to its suitability for any general or specific application.

The contents of this publication are not intended to be and should not be construed as a standard of the American Academy of Environmental Engineers® or of any of the associated organizations mentioned in this publication and are not intended for use as a reference in purchase specifications, contracts, regulations, statutes, or any other legal document.

No reference made in this publication to any specific method, product, process, or service constitutes or implies an endorsement, recommendation, or warranty thereof by the American Academy of Environmental Engineers® or any such associated organization.

Neither the American Academy of Environmental Engineers® nor any of such associated organizations or authors makes any representation or warranty of any kind, whether express or implied, concerning the accuracy, suitability, or utility of any information published herein and neither the American Academy of Environmental Engineers® nor any such associated organization or author shall be responsible for any errors, omissions, or damages arising out of use of this information.

CONTRIBUTORS

This monograph was prepared under the supervision of the WASTECH® Steering Committee. The manuscript for the monograph was written by a task group of experts in thermal desorption and was, in turn, subjected to two peer reviews. One review was conducted under the auspices of the Steering Committee and the second by professional and technical organizations having substantial interest in the subject.

PRINCIPAL AUTHORS

JoAnn Lighty, Ph.D., *Task Group Chair*
Assistant Professor
Department of Chemical and Fuel Engineering
University of Utah

Martha Choroszy-Marshall
CIBA-GEIGY
Program Manager
Thermal Treatment

Vic Cundy, Ph.D.
Professor & Chair
Mechanical Engineering Department
Louisiana State University

Michael Cosmos
Project Director
Roy F. Weston

Paul De Percin
Chemical Engineer
U.S. Environmental Protection Agency

In addition, Mr. Charles A. Cook is credited with substantial contributions to the monograph manuscript. At the time of writing, Mr. Cook was a doctoral candidate in Mechanical Engineering at Louisiana State University under the supervision of Dr. Cundy.

REVIEWERS

The panel that reviewed the monograph under the auspices of the Project Steering Committee was composed of:

Charles O. Velzy, P.E., DEE, *Chair*
Private Consultant

Carl Swanstrom
Senior Project Manager
Chemical Waste Management, Inc.

Joseph W. Bozzelli, Ph.D.
Distinguished Professor
New Jersey Institute of Technology

William L. Troxler, P.E.
Vice President
Focus Environmental, Inc.

Peter J. Kroll, P.E.
Manager, Process Systems Engineering
ICF Kaiser Engineers, Inc.

Walter J. Weber, Jr., Ph.D., P.E., DEE
Earnest Boyce Distinguished Professor
University of Michigan

STEERING COMMITTEE

Frederick G. Pohland, Ph.D., P.E., DEE
Chair
Weidlein Professor of Environmental Engineering
University of Pittsburgh

William C. Anderson, P.E., DEE
Project Manager
Executive Director
American Academy of Environmental Engineers

Paul L. Busch, Ph.D., P.E., DEE
President and CEO
Malcolm Pirnie, Inc.
Representing, American Academy of Environmental Engineers

Richard A. Conway, P.E., DEE
Senior Corporate Fellow
Union Carbide Corporation
Chair, Environmental Engineering Committee
EPA Science Advisory Board

William D. Goins, P.E.
Deputy Director for Technology Development
Office of the Secretary of Defense
U.S. Department of Defense

Timothy B. Holbrook, P.E.
District Engineering Manager
Groundwater Technology
Representing, Air and Waste Management Association

Walter W. Kovalick, Jr., Ph.D.
Director, Technology Innovation Office
Office of Solid Waste and Emergency Response
U.S. Environmental Protection Agency

Joseph F. Lagnese, Jr., P.E., DEE
Private Consultant
Representing, Water Environment Federation

Peter B. Lederman, Ph.D., P.E., DEE, P.P.
Center for Env. Engineering & Science
New Jersey Institute of Technology
Representing American Institute of Chemical Engineers

Raymond C. Loehr, Ph.D., P.E., DEE
H.M. Alharthy Centennial Chair and Professor
Civil Engineering Department
University of Texas

Timothy Oppelt, Ph.D.
Director, Risk Reduction Engineering Laboratory
U.S. Environmental Protection Agency

George Pierce, Ph.D.
Editor in Chief
Journal of Microbiology
Manager, Bioremediation Technology Dev.
American Cyanamid Company
Representing the Society of Industrial Microbiology

H. Gerard Schwartz, Jr., Ph.D., P.E.
Senior Vice President
Sverdrup
Representing, American Society of Civil Engineers

Claire H. Sink
Acting Director
Division of Technical Innovation
Office of Technical Integration
Environmental Education Development
U.S. Department of Energy

Peter W. Tunnicliffe, P.E., DEE
Senior Vice President
Camp Dresser & McKee, Incorporated
Representing, Hazardous Waste Action Coalition

Charles O. Velzy, P.E., DEE
Private Consultant
Representing, American Society of Mechanical Engineers

William A. Wallace
Vice President, Hazardous Waste Management
CH2M Hill
Representing, Hazardous Waste Action Coalition

Walter J. Weber, Jr., Ph.D., P.E., DEE
Earnest Boyce Distinguished Professor
University of Michigan

REVIEWING ORGANIZATIONS

The following organizations contributed to the monograph's review and acceptance by the professional community. The review process employed by each organization is described in its acceptance statement. Individual reviewers are, or are not, listed according to the instructions of each organization.

Air & Waste Management Association

The Air & Waste Management Association is a nonprofit technical and educational organization with more than 14,000 members in more than fifty countries. Founded in 1907, the Association provides a neutral forum where all viewpoints of an environmental management issue (technical, scientific, economic, social, political, and public health) receive equal consideration.

This worldwide network represents many disciplines: physical and social sciences, health and medicine, engineering, law, and management. The Association serves it's membership by promoting environmental responsibility and providing technical and managerial leadership in the fields of air and waste management. Dedication to these objectives enables the Association to work towards it's goal: a cleaner environment.

Qualified reviewers were recruited from the Technical Council Committee, Waste Division. It was determined that the monograph is technically sound and publication is endorsed.

The reviewers were:

James R. Donnelly

R. Sahu, Ph.D.

Wileen Sweet Dodge

American Institute of Chemical Engineers

The Environmental Division of the American Institute of Chemical Engineers has enlisted its members to review the monograph. Based on that review the Environmental Division endorses the publication of the monograph.

American Society of Mechanical Engineers

Founded in 1880, The American Society of Mechanical Engineers (ASME) is a nonprofit educational and technical organization, having at the date of publication of this document approximately 116,400 members, including 19,200 students. Members work in industry, government, academia, and consulting. The Society has thirty-seven technical divisions, four institutes, and three interdisciplinary programs which conduct more than thirty national and international conferences each year.

This document was reviewed by volunteer members of the Monograph Review Committee of the Solid Waste Processing Division and the Hazardous Waste Committee of the Safety Engineering and Risk Assessment Division of ASME, each with technical expertise and interest in the field covered by the document. Although, as indicated on the reverse of the title page of this document, neither ASME nor any of its Divisions or Committees endorses or recommends, or makes any representation or warranty with respect to, this document, those Divisions and Committees which conducted a review believe, based upon such review, that this document and the findings expressed are technically sound.

Hazardous Waste Action Coalition

The Hazardous Waste Action Coalition (HWAC) is a nonprofit association of engineering and science firms that provide hazardous waste remediation services for both public and private sector clients. Coalition member firms employ experts in over ninety technical disciplines, including all engineering disciplines.

Qualified reviewers were recruited from HWAC's Technical Practices Committee. After consulting with HWAC's lead reviewer, it was determined that the monograph is technically sound and publication is endorsed.

The reviewers were:

Kris Krishnaswami
Senior Associate
Malcolm Pirnie, Inc.

Robert G. Wilbourn
Process Development Manager
IT Corporation

Gil M. Zemansky, Ph.D., P.HGW.
Principal Hydrogeologist
Terracon Environmental, Inc.

Water Environment Federation

The Water Environment Federation is a nonprofit, educational organization composed of member and affiliated associations throughout the world. Since 1928, the Federation has represented water quality specialists including engineers, scientists, government officials, industrial and municipal treatment plant operators, chemists, students, academics, and equipment manufacturers, and distributors.

Qualified reviewers were recruited from the Federation's Industrial and Hazardous Wastes Committees. It has been determined that the monograph is technically sound and publication is endorsed.

The reviewers were:

Peter J. Cagnetta
Project Soil Scientist
R.E. Wright Associates, Inc.

Larry J. DeFlui
Project Environmental Engineer
R.E. Wright Associates, Inc.

Gomes Ganapathi
Section Manager
Waste Management Technologies Section
Science Application International Corporation

Stephen Gelman
District Operations Manager
CH2M Hill, Inc.

S. Bijoy Ghosh, P.E.*
Principal Engineer
Engineering-Science, Inc.

Michael Joyce
Director of Engineering Sales
R.E. Wright Associates, Inc.

Les Porterfield
Assistant Vice President
BCM Engineers, Inc.

Delmar H. Prah
Project Engineer
Argonne National Laboratory

Robert C. Williams, P.E., DEE
Director of the Division of Health Assessment and Consultation
Agency for Toxic Substances and Disease Registry

* WEF lead reviewer

ACKNOWLEDGMENTS

The WASTECH® project was conducted under a cooperative agreement between the American Academy of Environmental Engineers® and the Office of Solid Waste and Emergency Response,U.S. Environmental Protection Agency. The substantial assistance of the staff of the Technology Innovation Office was invaluable.

Financial support was provided by the U.S. Environmental Protection Agency, Department of Defense, Department of Energy, and the American Academy of Environmental Engineers®.

This multiorganization effort involving a large number of diverse professionals and substantial effort in coordinating meetings, facilitating communications, and editing and preparing multiple drafts was made possible by a dedicated staff provided by the American Academy of Environmental Engineers® consisting of:

Paul F. Peters
Assistant Project Manager & Managing Editor

Susan C. Richards
Project Staff Assistant

J. Sammi Olmo
Project Administrative Manager

Yolanda Y. Moulden
Staff Assistant

I. Patricia Violette
Staff Assistant

TABLE OF CONTENTS

Appendices

LIST OF TABLES

LIST OF FIGURES

1 INTRODUCTION

This monograph on thermal desorption is one of a series of eight on innovative site and waste remediation technologies that are the culmination of a multiorganization effort involving more than 100 experts over a two-year period. It provides the experienced, practicing professional guidance on the application of innovative processes considered ready for full-scale application. Other monographs in this series address bioremediation, chemical treatment, soil washing/soil flushing, solvent/chemical extraction, stabilization/solidification, thermal destruction, and vacuum vapor extraction.

1.1 Thermal Desorption

The thermal desorption processes addressed in this monograph use heat, either direct or indirect, ex situ, as the principal means to physically separate and transfer contaminants from soils, sediments, sludges, filter cakes, or other media. Thermal desorption is part of a treatment train; some pre- and postprocessing is necessary. Thermal desorbers are physical separation facilities and are not specifically designed to decompose organics (organic denotes compounds, including volatiles, semivolatiles, PCBs, and pesticides); depending on the organics present, and the temperature of the system, however, some decomposition may occur.

The separated contaminants, water vapor, and particulates must be collected and treated. This is typically accomplished using conventional methods of condensation, adsorption, incineration, filtration, and the like. The methods selected depend on the nature and concentration of contaminants, regulations, and the economics of the systems employed. It may be possible to reuse the treated material, and, in some cases, the recovered contaminants may have commercial value. Regulations that govern thermal destruction processes may apply in some cases to some thermal desorption processes.

The effectiveness of the desorber is generally measured by comparing the contamination levels in the media pre- and posttreatment.

1.2 *Development of the Monograph*

1.2.1 Background

Acting upon its commitment to develop innovative treatment technologies for the remediation of hazardous waste sites and contaminated soils and ground water, the U.S. Environmental Protection Agency (EPA) established the Technology Innovation Office (TIO) in the Office of Solid Waste and Emergency Response in March, 1990. The mission assigned TIO was to foster greater use of innovative technologies.

In October of that same year, TIO, in conjunction with the National Advisory Council on Environmental Policy and Technology (NACEPT), convened a workshop for representatives of consulting engineering firms, professional societies, research organizations, and state agencies involved in remediation. The workshop focused on defining the barriers that were impeding the application of innovative technologies in site remediation projects. One of the major impediments identified was the lack of reliable data on the performance, design parameters, and costs of innovative processes.

The need for reliable information led TIO to approach the American Academy of Environmental Engineers. The Academy is a long-standing, multidisciplinary environmental engineering professional society with wide-ranging affiliations with the remediation and waste treatment professional communities. By June 1991, an agreement in principle (later formalized as a Cooperative Agreement) was reached. The Academy would manage a project to develop monographs describing the state of available innovative remediation technologies. Financial support would be provided by the EPA, U.S. Department of Defense (DOD), U.S. Department of Energy (DOE), and the Academy. The goal of both TIO and the Academy was to develop monographs providing reliable data that would be broadly recognized and accepted by the professional community, thereby, eliminating or, at least, minimizing this impediment to the use of innovative technologies.

The Academy's strategy for achieving the goal was founded on a multiorganization effort, WASTECH® (pronounced Waste Tech), which joined in partnership the Air and Waste Management Association, the American Institute of Chemical Engineers, the American Society of Civil Engineers, the American Society of Mechanical Engineers, the Hazardous Waste Action Coalition, the Society for Industrial Microbiology, and the Water Environment Federation, together with the Academy, EPA, DOD, and DOE. A Steering Committee composed of highly respected representatives of these organizations having expertise in remediation technology formulated the specific project objectives and process for developing the monographs (See page iv. for a listing of Steering Committee members).

By the end of 1991, the Steering Committee had organized the Project. Preparation of the monograph began in earnest in January, 1992.

1.2.2 Process

The Steering Committee decided upon the technologies, or technological areas, to be covered by each monograph, the monographs' general scope, and the process for their development and appointed a task group composed of five or more experts to write a manuscript for each monograph. The task groups were appointed with a view to balancing the interests of the groups principally concerned with the application of innovative site and waste remediation technologies — industry, consulting engineers, research, academe, and government (see page iii for a listing of members of the Thermal Desorption Task Group).

The Steering Committee called upon the task groups to examine and analyze all pertinent information available, within the Project's financial and time constraints. This included, but was not limited to, the comprehensive data on remediation technologies compiled by EPA, the store of information possessed by the task groups' members, that of other experts willing to voluntarily contribute their knowledge, and information supplied by process vendors.

To develop broad, consensus-based monographs, the Steering Committee prescribed a twofold peer review of the first drafts. One review was conducted by the Steering Committee itself, employing panels consisting of two members of the Committee supplemented by at least four other experts (See *Reviewers*, page iii, for the panel that reviewed this monograph). Simultaneous with the Steering Committee's review, each of the professional and technical organizations represented in the Project reviewed those monographs addressing technologies in which it has substantial interest and competence. Aided by a Symposium sponsored by the Academy in October 1992, persons having interest in the technologies were encouraged to participate in the organizations' review.

Comments resulting from both reviews were considered by the Task Group, appropriate adjustments were made, and a second draft published. The second draft was accepted by the Steering committee and participating organizations. The statements of the organizations that formally reviewed this monograph are presented under *Reviewing Organizations* on page v.

1.3 Purpose

The purpose of this monograph is to further the use of innovative thermal desorption site remediation and waste processing technologies, that is, technologies not commonly applied, where their use can provide better, more cost-effective performance than conventional methods. To this end, the monograph documents the current state of a number of innovative thermal desorption technologies.

1.4 Objectives

The monograph's principal objective is to furnish guidance for experienced, practicing professionals and users' project managers. The monograph is intended, therefore, not to be prescriptive, but supportive. It is intended to aid experienced professionals in applying their judgment in deciding whether and how to apply the technologies addressed under the particular circumstances confronted.

In addition, the monograph is intended to inform regulatory agency personnel and the public about the conditions under which the processes it addresses are potentially applicable.

1.5 Scope

The monograph addresses innovative thermal desorption technologies which are not yet conventional, that is, not commonly applied, that have been sufficiently developed so that they can be used in full-scale applications. It addresses all such technologies for which sufficient data was available to the Thermal Desorption Task Group to describe and explain the technology and assess its effectiveness, limitations, and potential applications. Laboratory- and pilot-scale technologies were addressed, as appropriate.

The monograph's primary focus is site remediation and waste treatment. To the extent the information provided can also be applied to production waste streams, it will provide the profession and users this additional benefit. The monograph considers all waste matrices to which thermal desorption processes can be reasonably applied, such as, soils, sludges, filter cakes, and other solid media.

Application of site remediation and waste treatment technology is site specific and involves consideration of a number of matters besides alternative technologies. Among them are the following that are addressed only to the extent essential to understand the applications and limitations of the technologies described:

- site investigations and assessments;
- planning, management, specifications, and procurement;
- regulatory requirements; and
- community acceptance of the technology.

1.6 Limitations

The information presented in this monograph has been prepared in accordance with generally recognized engineering principles and practices and is for general information only. This information should not be used without first securing competent advice with respect to its suitability for any general or specific application.

Readers are cautioned that the information presented is that which was generally available during the period when the monograph was prepared. Development of innovative site remediation and waste treatment technologies is ongoing. Accordingly, postpublication information may amplify, alter, or render obsolete the information about the processes addressed.

This monograph is not intended to be and should not be construed as a standard of any of the organizations associated with the WASTECH® Project; nor does reference in this publication to any specific method, product, process, or service constitute or imply an endorsement, recommendation, or warranty thereof.

1.7 Organization

This monograph and others in the series are organized under a uniform outline intended to facilitate cross reference among them and comparison of the technologies they address. Chapter 2.0, Process Summary, provides an overview of all material presented. Chapter 3.0, Process Identification and Description, provides comprehensive information on the processes addressed. Each process is fully analyzed in turn. The analysis includes a description of the process (what it does and how it does it), its scientific basis, status of development, environmental effects, pre- and posttreatment requirements, health and safety considerations, design data, operational considerations, and comparative cost data to the extent available. Also addressed are process unique planning and management requirements and process variations.

Chapter 4.0, Potential Applications, Chapter 5.0, Process Evaluation, and Chapter 6.0, Limitations, provide a synthesis of available information and informed judgments on the processes. Each of these chapters addresses the processes in the same order as they are described in Chapter 3.0. Technology Prognosis, Chapter 7.0, identifies other processes or elements of processes that require further research and demonstration before full-scale application can be considered.

2
PROCESS SUMMARY[1]

2.1 *Process Identification and Description*

Thermal desorption is an ex situ means for physically separating organics from soils, sediments, sludges, filter cakes, and other solid media. Thermal desorbers are not specifically designed to effect decomposition. Desorber performance is generally measured by comparing the contaminant levels in the untreated medium with the contamination levels remaining in the processed medium.

The contaminated material is excavated and delivered to the thermal desorber. Typically, large objects are screened from the medium and rejected. Rejected material can sometimes be sized and recycled to the desorber feed. The medium is then delivered by gravity to the desorber inlet or conveyed by augers to a feed hopper from which it is mechanically conveyed to the desorber.

There are two approaches to thermal desorption remediation: stationary facilities to which the contaminated media is transported and mobile systems that operate on site. Both kinds of facilities are available for treating petroleum-contaminated wastes; however, only mobile systems are available for treating Comprehensive Environmental Response, Compensation, and Liability Act (CERCLA) wastes.

There are significant variations in the desorption step reflected in the following classification of thermal desorption systems:

- direct-fired rotary desorbers;
- indirect-fired rotary desorbers;

1. This chapter is a summary of Chapters 3.0 through 7.0. Sources are cited, where appropriate, in those chapters — Ed.

- direct- or indirect-heated conveyor systems; and
- SoilTech System.

Limited data are available on fluidized beds and one other desorption system that cannot be placed in any of the above classes, the Texarome Process. Therefore, these systems are not addressed in the monograph proper but are described in Appendix A. See figure 2.1 for a schematic of an example of a thermal desorption system. See also tables 2.1 (on page 2.3), 2.2 (on page 2.4), and 2.3 (on page 2.5) for summary lists of the key features of the equipment of each of the three major classes of thermal desorption systems (the first three listed above) to facilitate a general comparison.

In a desorption unit, heat is transferred to the solid media. The contaminated material is heated, and water and the contaminants are devolatilized. An inert gas, such as nitrogen or oxygen-deficient (less than 4%) combustion offgas, may be injected as a sweep stream. Organics in the offgas may be collected and recovered by condensation and adsorption or burned in an afterburner. Particu-

Figure 2.1

Treatment System Schematic

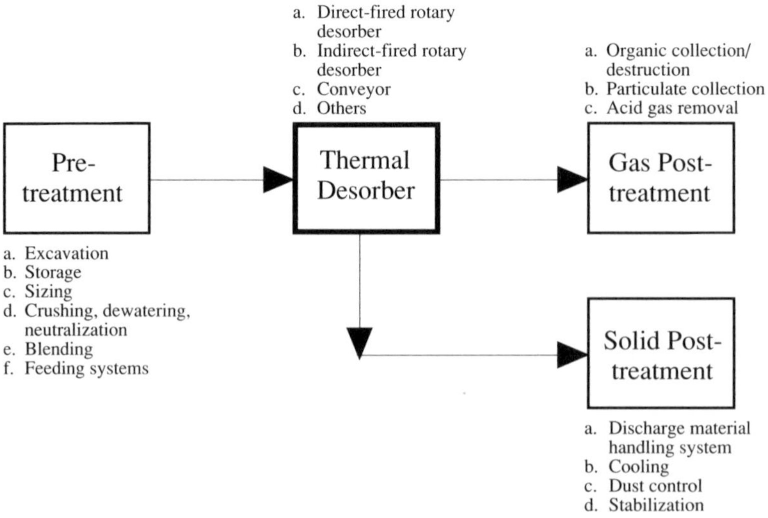

late matter is removed by conventional air pollution control (APC) methods. The selection of the gas treatment system will depend on the concentrations of the contaminants, cleanup standards and regulations, and the economics of the offgas treatment system(s) employed.

Operation of thermal desorption systems can create a number of process residual streams: treated media; untreated, oversized rejects; condensed contaminants and water; particulate control-system dust; clean offgas; and spent

Table 2.1

Design and Operating Characteristics

System	Heat	Operating Parameters	Other
Rotary Drum — Direct fired	■ Duties from 7-100 MM Btu/hr ■ 25,000 Btu/hr per ft ^{3} internal volume ■ Propane, natural gas, or fuel oil ■ Kiln materials Carbon Steel: up to 315°C Alloy Steel: up to 650°C	■ Rotational speeds 0.25 to 10 rpm ■ L/D: 2:1 to 10:1 ■ Soil residence time to 10 min for petroleum, 20 to 30 for SVOC ■ Feed rates: up to 40 ton/hr1	■ Co- or countercurrent gas flow ■ Negative pressure in the dryer ■ Variations in gas clean up ■ Need to be careful not to exceed LEL ■ Fire hazards (baghouse) and explosion (storage building)
Rotary Drum — Indirect fired	■ Propane or natural gas ■ Kiln materials Carbon Steel: up to 315°C Alloy Steel: up to 600°C ■ 40% of input heat is transferred to waste material	■ Rotational speeds up to 2.5 rpm ■ L/D: 5:1 to 10:1 ■ Feed rate up to 10 ton/hr	■ Negative pressure in the dryer ■ Variations in gas cleanup (usually smaller volume) ■ Same hazards as above
Heated Conveyers	■ Discharge temperature dependent on heating media; up to 370°C maximum ■ Indirect or directly heated	■ Soil residence time up to 90 min ■ Feed rate: up to 10 ton/hr	■ Same hazards as above
SoilTech ATP	■ Natural gas or propane ■ Up to 590°C	■ Feed rate: 10 ton/hr (25 ton/hr in design)	■ 4 zones of physical processes

All units are available, either stationary or mobile
Treatability studies are required

carbon, if used. Treated media, debris, and oversized rejects may be suitable for return on site.

Treated condensed water and/or treated scrubber purge water (blowdown) can be purified and returned to the site wastewater treatment facility (if available), disposed to a sewer system, or used for rehumidification and cooling of the hot dusty media. If produced, concentrated, condensed organic contaminants are stored for shipment to recycling centers or off-site treatment facilities, such as incinerators.

Table 2.2

Pretreatment Requirements

- Storage Required
- Size Distribution:
  - Generally less than 2 to 2.5 in. (see text)
  - Crush and/or screen large solids
  - May need to remove magnetic materials
  - Special requirements for asphalt production
- Contaminant Characterization
  - Define contaminants and cleanup criteria for system
  - For direct-fired units operation, Btu content must be below limits
  - May require blending for homogeneity
  - Generally pH should be above 5 (and less than 11)
- Moisture Content
  - Dewater to 20 to 50% before feeding
  - Thaw frozen soils if necessary

Dust collected from particulate control devices may be combined with the treated medium or, depending on the results of contaminant analyses, recycled through the desorption unit.

Clean offgas is usually released to the atmosphere, although systems that use inert gas, for example, nitrogen, recycle the gas to the desorber after treatment. Activated carbon can also be used to treat both the gases and condensed water, and both on and off-site regeneration of activated carbon could be used.

Environmental impacts associated with all thermal desorbers, aside from process emissions, are attributable to excavation of contaminated solids, management of treated solids, and equipment noise. Appropriate material handling measures are needed to control fugitive emissions of dust and highly volatile contaminants following excavation and before processing. Treated solids should be cooled and stored while testing takes place, which can take several days. Accordingly, dust, runoff, and runon controls are needed.

It is not possible to differentiate among the thermal desorbers based on cost. The costs are scale dependent, ranging from $90-130/ton for a 1,000 ton site to $40-70/ton for a 10,000 ton site for mobile systems treating petroleum-contaminated soils and from $300-600/ton for a 1,000 ton site to $150-200/ton for a 10,000 ton site mobile system operating at a CERCLA site. Matrix moisture and contaminant type are critical parameters in analyzing desorption costs. Cost for treating petroleum-contaminated soils in stationary facilities may be as low as $35 per ton.

Table 2.3

Posttreatment Requirements

System	Treated Soil	Gas Treatment	Liquid Treatment
Rotary Drum — Direct fired	▪ Water quench ▪ Stabilization for heavy metals ▪ Disposal	▪ Direct fired require larger capacity ▪ Organic control: afterburner, catalytic oxidizer, carbon adsorption ▪ Acid gas and particulate removal: venturi scrubber, acid neutralization, cyclone, baghouse	▪ Treat water for organics using carbon adsorption ▪ Filter to remove solids ▪ NPDES or POTW requirements apply ▪ If there are organic liquids, dispose
Rotary Drum — Indirect fired	▪ Same as above	▪ Indirect: lower gas volume ▪ Organic control: afterburner, condenser, carbon adsorption ▪ Acid gas and particulate removal: same as above, but not required in all units	▪ Same as above
Heated Conveyers	▪ Same as above	▪ Same as above depending on direct or indirect fired	▪ Same as above

2.2 Potential Applications

Thermal desorption technology appears to be applicable to many types of waste streams. As of February 1992, thermal desorption had been chosen for 28 remedial actions whose projects were in various stages from predesign through completion (US EPA 1992e). Effective removal of a number of contaminants has been demonstrated, including those in petroleum-contaminated and PCB-contaminated soils, volatile and semi-volatile organics, pesticides, and manufactured gas plant soils containing hydrocarbons. The process generates some residual streams that must disposed of either off or on site.

Thermal desorption is generally not effective in separating inorganics from the contaminated medium. Very volatile metals, such as, mercury, however, can be removed by these processes.

2.3 Process Evaluation

Thermal desorption has been proven effective in removing organics to levels in established cleanup standards from contaminated soils, sludges, sediments, and filter cakes. Chemical contaminants for which bench-scale through full-scale treatment data exist include volatile organic compounds (VOCs), semivolatile organic compounds (SVOCs), polynuclear aromatic hydrocarbons (PAHs or PNAs), polychlorinated biphenyls (PCBs), pesticides, and dioxins and furans. Volatile organic compounds have been commonly targeted at CERCLA sites where thermal desorption was selected; VOCs were targeted at 19 sites. Polychlorinated biphenyls were targeted at five sites, other SVOCs at five sites, and pesticides at three sites (US EPA 1992e). While some mixed wastes (radioactive and hazardous) have also been treated using thermal desorption, this application of the technology will not be addressed in this monograph.

There are more than 150 full-scale thermal desorbers available in the U.S. (Troxler et al. 1992) and in operation. This number includes asphalt aggregate dryers for remediating petroleum-contaminated soils. Although most of these units treat only petroleum-contaminated soils, the use of this approach for Superfund site remediation is increasing. The number of times that thermal desorption has been selected in a Record of Decision (ROD) as the remediation method for Superfund sites has grown from one in 1985 to ten in 1991 (US EPA 1992e). Thermal desorption appears in the RODs for 28 Superfund sites as of February 1992 (US EPA 1992e).

The cost of thermal desorption treatment is principally a function of the solid moisture content, solid characteristics, contaminant volatility, contaminant concentration, vendor equipment limitations, and cleanup standards. Regulatory requirements may be a key contributor to the cost of treatment and those that govern thermal destruction processes may apply to certain thermal desorption systems.

2.4 *Limitations*

The primary technical factor affecting thermal desorption performance is the maximum bed temperature effected in the solid media. Since the basis of the process is physical removal of contaminants from the medium by volatilization, bed temperature largely determines the type of organics that will be removed and the effectiveness of removal.

Material handling of soils that are tightly aggregated, such as clays, certain rock fragments, or particles greater than 2.5 to 5.0 cm (1 to 2 in.), can result in poor processing performance because of caking. Caking occurs if soil moisture is above the plastic limit. Also, if a high fraction of fine silt or clay exists in the matrix, fugitive dusts will be generated and a greater dust loading will be placed on the downstream air pollution control equipment. The treated medium will typically contain less than 1% moisture. Dust can easily form in the transfer of the treated medium from the desorption unit, but can be mitigated by water sprays. Desorption systems that produce a condensed water stream normally use it for wetting the treated material.

There is evidence that with some system configurations some materials, such as tars, may foul and/or plug heat transfer surfaces. Both laboratory and field tests have documented the deposition of insoluble brown tars on internal system components.

There are also limitations as to the concentration of organic contaminants that can be thermally treated in any one process. First, with regard to excavation of the site, as with any ex situ technology, concentrations must be such that fugitive emissions are not excessive. In addition, vapor organic concentrations within the thermal desorber must be kept below 25% of the lower explosive limits (LEL) if the desorber is operated with excess oxygen.

2.5 Technology Prognosis

This technology continues to evolve, particularly as applied in hazardous waste site remediation. More information is needed concerning the scaling of laboratory results with full-scale systems, the fate of metals in desorbers, and the formation of dioxins. The emission of metals, especially volatile metals, such as mercury, needs to be understood. In addition, the degree of leachability of metals in the ash needs to be determined for purposes of ultimate disposal.

3

PROCESS IDENTIFICATION AND DESCRIPTION

Of the 76 demonstrations being conducted under the United States Environmental Protection Agency (US EPA) Superfund Innovative Technology Evaluation (SITE) Program, 17% of the technologies being evaluated are thermal desorption processes (US EPA 1991a). Through fiscal year 1991, of 498 remedial actions, thermal desorption was selected as the alternative technology for 28. Alternative technology accounted for 210 (42%) of the treatment technologies selected (US EPA 1992e).

3.1 *Description*

The thermal desorber is one part of the total system used in the remediation of contaminated solid media (see figure 2.1 on page 2.2). Before thermal desorption, excavation and pretreatment — material handling, material sizing, and removal of large objects — will be required. There are significant variations in the desorption step reflected in distinct classes of thermal desorber systems. In this monograph, the systems are classified and addressed as:

- direct-fired rotary desorbers;
- indirect-fired rotary desorbers; and
- direct or indirect-heated conveyor systems.

Another rotary desorber system, the SoilTech process, could not be classified exclusively as indirect or direct and, therefore, is addressed here as a separate class. Two other systems, fluidized beds and the Texarome Process, were identified, but insufficient data were available to enable the authors to address them at the time of this writing (August 1992). They are briefly described in Appendix A.

Following the thermal desorption step, posttreatment is usually necessary. Posttreatment consists of: gas treatment (through condensation units, afterburners, carbon adsorption units), solids' treatment (quenching, stabilization, disposal), and liquid treatment (water treatment, organic liquid treatment, and disposal in the case of condensing system). The objectives of the overall treatment system are clean solids, environmentally acceptable stack gases and water, and complete disposal of all other residuals.

See also the US EPA Superfund *Engineering Bulletin: Thermal Desorption Treatment*, of April 1993, appended hereto as Appendix B.

3.2 **Scientific Basis**

In any thermal desorption system, heat must be transferred to the solid particles to vaporize the contaminants from particles; in turn, the vaporized contaminants must be transferred from the particles to the gas phase. The specific modes of heat and mass transfer vary with the type of thermal desorption system employed. Figure 3.1 (on page 3.3) depicts the transport mechanisms to be considered. As explained in Owens et al. (1991), system temperature will determine the importance of radiative heat transfer. Interparticle phenomena refer to heat and mass transfer within the bed. Heat and mass transfer processes at the interparticle level are distinct, depending on the thermal desorption system used. For example, in a rotary desorber, heat must be transferred to the bed of solids by radiative, convective, and conductive heat transfer with the wall and gas, while mass must be transferred through the bed of solids. At the gas/solid interface, mass must then move into the free stream gas.

Intraparticle processes, referring to transfer of heat and mass between the particle and the bulk environment within the bed, are also important. These processes are not dependent on the system, since they are fundamental to the contaminant and type of media being treated. The work of Keyes (1992) showed that in toluene desorption from montmorillonite clay, local equilibrium exists within pores and that the effective rate of desorption from individual particles is controlled by intraparticle diffusion. Bozzelli and coworkers (Wu, Dong, and Bozzelli 1992; Wu and Bozzelli 1992) found that, when assuming linear equilibrium (where the concentration in the soil is directly proportional to the gas-phase concentration) within the pores, the equilibrium constants were

Figure 3.1

Schematic of the Transport Phenomena Occurring During Thermal Treatment of a Solid Bed

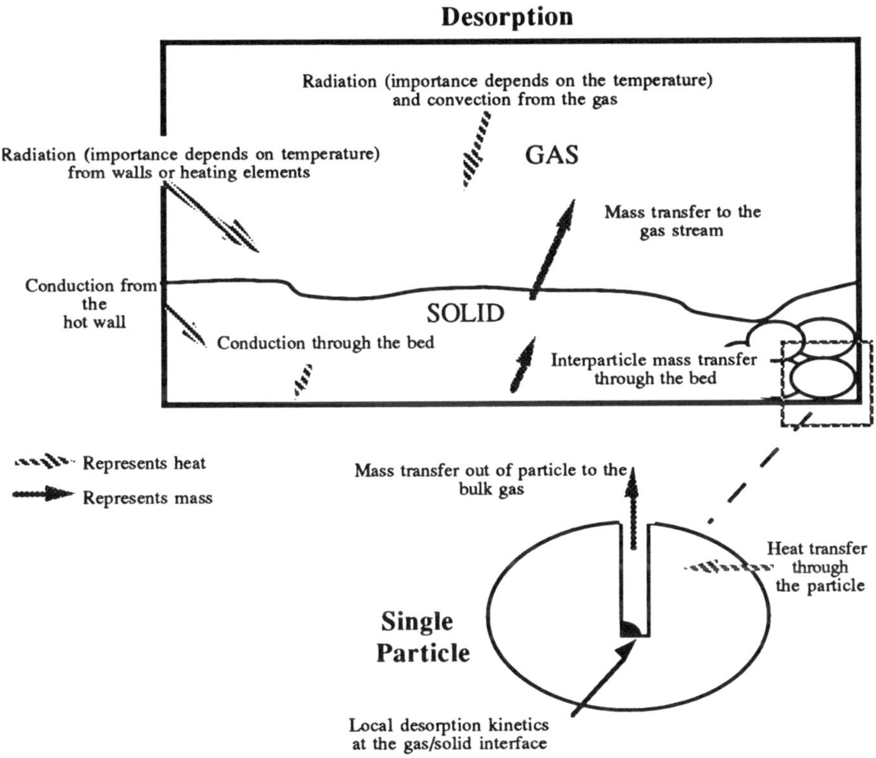

Included are heat transfer (conduction, radiation) and mass transfer (from particle, from bed) processes.

strongly dependent upon temperature for their soil/contaminant systems. In addition, they found no effect of particle size on the equilibrium constants. Keyes showed also, however, that when these particles constitute a sorbent bed in a rotary kiln, mass-transport resistances associated with the sorbent bed control the overall desorption rate and that equilibrium exists between the adsorbed contaminant and the interparticle gas phase. The concept of local equilibrium within pores is supported by the work of Gorte (1982), Herz, Kiela, and Marin (1982), and Jones and Griffin (1983).

Researchers have identified several important variables that need to be considered in the equilibrium between contaminants and soil particles. They have demonstrated that contaminant removal is highly dependent on the following parameters:

- Temperature — modest increases in temperature greatly decrease residual concentrations; for example, Helsel and Groen (1988) found that at 300°C (570°F) the Pyrene residual was 1.2% of the original. At 400°C (750°F), this dropped to 0.01%. The tests were bench-scale at an initial concentration of 1,400 ppm;
- Soil matrix — coarse particles such as sands will desorb contaminants easier than fine grained clays and silts;
- Contaminant — some contaminants will bind strongly to soils while others will not; and
- Moisture content — increased moisture reduces the capacity of the contaminant to adsorb on soils with high mineral contents (silts and clays).

(Varuntanya et al. 1989; Flytzani-Stephanopoulos et al. 1991; Lighty et al. 1988; Lighty, Silcox et al. 1990; Helsel and Groen 1988; Rogers, Holsen, and Anderson 1990; and Gilot et al. 1992.)

Researchers have found that, while the initial 90% of a contaminant might be easily removed, the final 10% will take much longer, especially if the cleanup criteria is in the parts per billion range. See figure 3.2 (on page 3.5). This phenomena is due to the adsorptive properties of the soil, which may have a tendency to strongly "hold" monolayers (single molecules) of contaminant to its surface. Lighty, Silcox et al. (1990), Locke, Arozarena, and Chambers (1991), and Borkent-Verhage et al. (1986) found that the relationship between temperature and removal of contaminant is nonlinear, confirming the possibility of a monolayer effect.

Limited performance data for thermal desorption systems will be presented in the subsequent chapters. For each remediation, however, treatability studies should be conducted, and data reviewed to determine the applicability of the technology (US EPA 1992d). Specifically, it is important to determine the solid

Figure 3.2

Relationship of Contaminant Removal Time for Increasing Temperatures

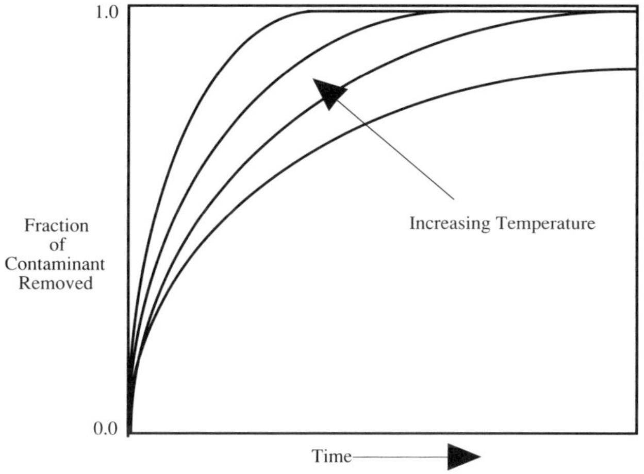

residence time and temperature required to meet the cleanup criteria. The studies should be conducted such that interparticle heat and mass transfer phenomena are minimized, since these processes are system specific.

3.3 *Waste Characterization*

See figure 3.3 (on page 3.6) for a summary of waste characteristics and related concerns. Waste characterization must be performed relative to cleanup criteria. It is important to understand not only the nature of the contaminants but also, where a solid is to be treated, the structure of the solid and the binding of the waste to the solid. See table 3.1 (on page 3.7) for analyses of the chemical and physical properties of the solid and contaminant that may need to be performed. Laboratories performing these analyses should meet US EPA ac-

Figure 3.3

Summary of Waste Characteristics and Related Concerns

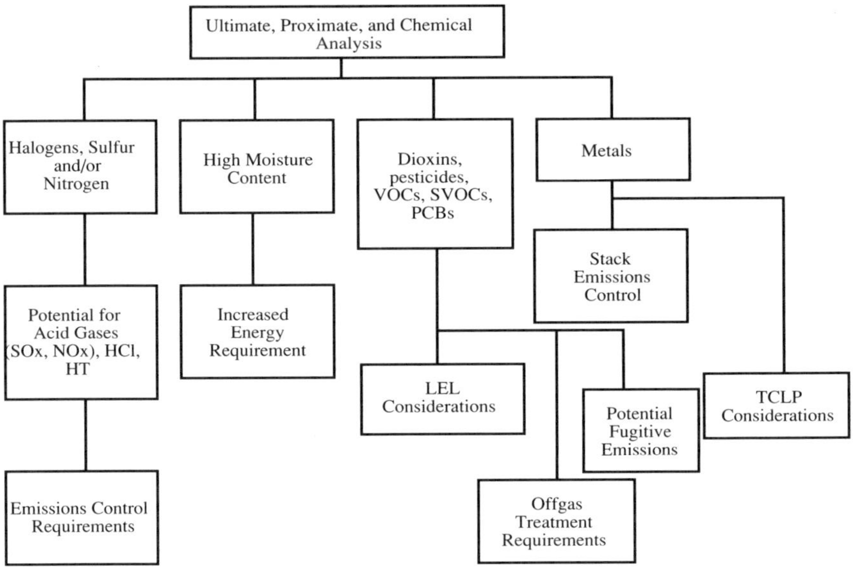

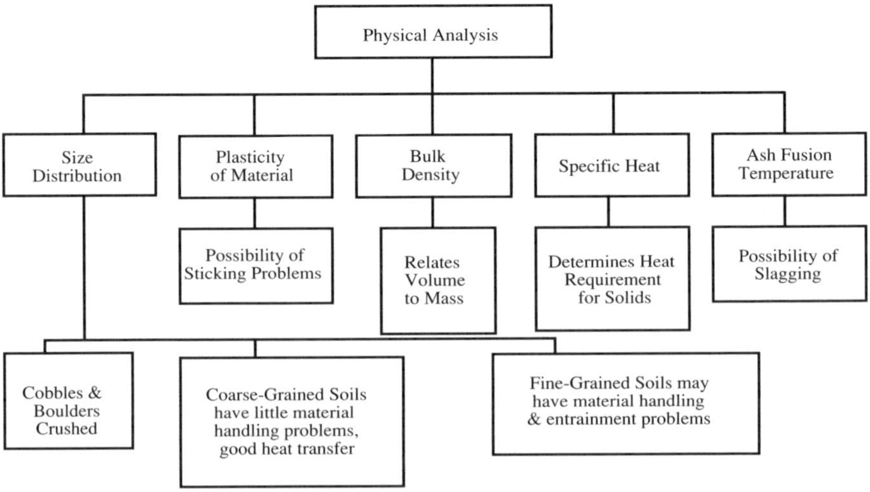

creditation requirements. Their quality assurance programs should meet prescribed standards, a priority requirement.

Sulfur and nitrogen are analyzed to determine whether there is potential for production of the pollutants sulfur dioxide and nitrogen oxides. Moisture content is important because of the energy required to heat and vaporize the water in the solid. Moisture is a major heat sink in the thermal desorber. See figure 3.4 (on page 3.8). Under the conditions specified therein, high-moisture content solids never rise above the vaporization temperature of the water because of the given heat input. In this case, it would be necessary to change the feed rate or firing rate in order to bring the material to the temperature.

Table 3.1

Waste Characterization Analyses

CHEMICAL	
Ultimate:	C, H, O, N, P and S (if waste has greater than 1% sulfur, then analyze for pyritic, SO_4 and organic sulfur)
Proximate:	moisture, ash, fixed carbon, volatiles
Halides:	Cl (total and ionic), F, I, Br
Organics:	Pesticides, dioxin, semivolatiles, volatiles, and PCBs
Metals:	Ag, As, Cd, Cr, Ba, Be, Hg, Pb, Se, Ni, Sb, Tl
PHYSICAL	
Bulk density	
Heating value	
Specific heat	
Ash Fusion Temperatures:	initial deformation, softening, hemispherical, fluid (if the ash fusion initial deformation temperature is less than 2,000°F, then also perform Na, K, and Ca)
pH (for corrosion considerations)	
Flash point	
Liquid limit	
Plasticity index	
Soil grain size	

Depending upon the site history, specific chemical evaluations and analyses of compounds such as dioxins, Polychlorinated Biphenyls (PCBs), pesticides, volatiles, and semivolatiles should be performed to determine the nature and concentration of contaminants in the waste. Knowledge of the initial concentration of contaminants is important for the following reasons:

- During excavation, fugitive emissions must be considered. To ensure safe operation, 25% of the lower explosion limit (LEL) should not be exceeded within a thermal desorber operating in an oxygen atmosphere;
- The types and amounts of organics that are being removed are factors in an engineering analysis of the offgas treatment systems;

Figure 3.4

Model Predictions Showing the Effect of the Initial Weight Fraction of Moisture in the Solid Feed on Bed Temperature Profile

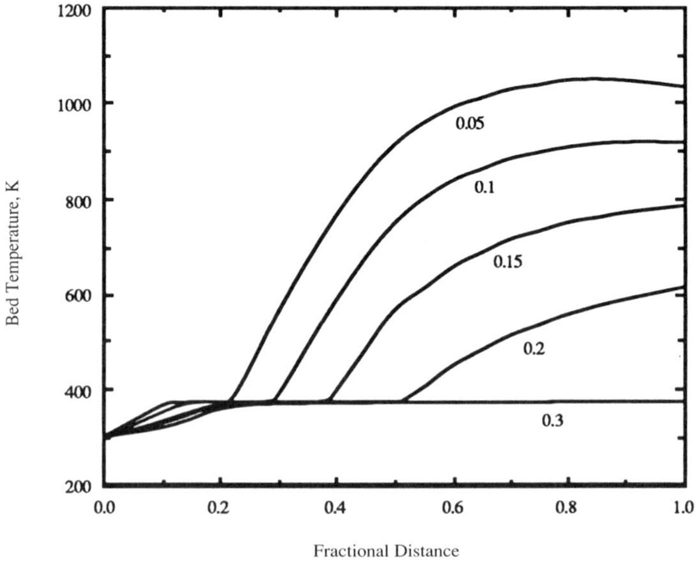

The dry solid feed rate is constant at 2.55 kg/sec.

Reprinted by permission of the Air and Waste Management Association from "The Effects of Rotary-Kiln Operating Conditions and Design on Burden Heating Rates as Determined by a Mathematical Model of Rotary Kiln Heat Transfer" by G.D. Silcox and D.W. Pershing, Journal of Air and Waste Management Association, Volume 40, March, 1990. Copyright 1990 by the Air and Waste Management Association.

- Heat release is the critical parameter for the design of an afterburner; and
- The composition of the organics is critical to the design of a condenser recovery system.

The residual material should be analyzed to verify that treatment standards have been met.

Analysis of concentrations of metals must be performed where volatile metals are present and might exit the stack. In assessments of hazardous waste incinerators, risk is almost always driven by the metals and dioxin emitted from the stack when indirect pathways are considered. Metals emissions from some incinerators, along with those from all boilers and industrial furnaces, are regulated under the Boiler and Industrial Furnace (BIF) Regulations (Appendix VIII of 40 CFR Part 261). Following are the metals whose emissions are regulated:

- Arsenic (As);
- Beryllium (Be);
- Cadmium (Cd);
- Chromium (Cr);
- Antimony (Sb);
- Barium (Ba);
- Mercury (Hg);
- Silver (Ag);
- Thallium (Tl); and
- Lead (Pb).

The first four are identified as carcinogens. The Resource Conservation Recovery Act (RCRA) lists, in addition, nickel (Ni), selenium (Se), and osmium (Os). Regulations under the RCRA and BIF, as well as state regulations, may govern emissions from thermal desorbers, depending on the regulatory status of the overall system (see also Section 6.8). The treatability test (see Subsection 4.1.2) will help determine the amount of vaporization that might occur at the desorber solid temperature.

The Toxic Characteristic Leachate Procedure (TCLP) metals analysis should be performed on the solid residuals to determine the proper disposal procedure. Residue that fails the TCLP test is precluded from land disposal under 40 CFR

Part 268 and, therefore, must be processed further. If the waste was classified as hazardous under the RCRA before being treated and passes the TCLP test, it may be placed in a RCRA landvault. If waste that was being remediated under the Comprehensive Environmental Response, Compensation, and Liability Act (CERCLA) passes the TCLP test, the treated waste may be placed back in the excavation hole, if allowed under the consent decree governing the remediation (see Section 6.8).

The physical properties of the solid must also be considered (US EPA 1992g). Knowledge of the particle size distribution of the solids to be remediated is important for proper selection of the type of thermal desorber and pretreatment scheme that will be used. Large boulders and cobbles (materials with equivalent diameters between 7.5 and 30 cm (3.0 and 12 in)) will need to be crushed or removed. Coarse-grained soils (gravels and sand that have more than 50% material retained on a 7.7-cm sieve) are generally free flowing and do not agglomerate into large particles. These solids have low moisture adsorption capacities and relatively good heat transfer characteristics. Silts, clays, organic soils, and peat are fine-grained and absorptive. Moisture content greatly affects the material handling of these soils. With fine-grained soils, entrained particulate can cause problems. Since the degree of particulate entrainment is directly related to gas-flow rate and particle size, different types of thermal desorber units will entrain varying quantities of solid.

The plasticity of the material is also important, because this characteristic will determine whether the material likely to stick to screening, sizing, conveying, and desorber equipment. Thermal treatment of a fine-grained soil with a moisture content above the plastic limit is extremely difficult (US EPA 1992g). Bulk density relates the volume of solid that needs to be remediated (the number usually given in the remedial investigation report) to the mass of solid that will be treated (a performance characteristic of a specific desorber system). The specific heat of the solid must be known to determine the amount of heat required to raise the temperature of the material. The ash-fusion temperatures are important in higher temperature environments, where the solid residue might form a slag. It is unlikely that desorbers will operate above 650°C (1,200°F); therefore, the ash-fusion temperature will not likely be exceeded. In systems with afterburners, however, slagging may present a problem because of the higher temperature environment.

3.4 *Rotary Desorber — Direct Fired*

3.4.1 Description

Direct-fired rotary desorber systems may be mobile or stationary. The typical system (see figure 3.5) consists of three components: the pretreatment and material handling systems, the desorption unit, and the posttreatment systems for both the gas and the solid.

3.4.1.1 Pretreatment

If storage of solids is required after excavation, the pretreatment process begins with proper fugitive emissions control and/or ventilation. The pretreatment process continues with screening to remove large material (to be crushed or manually cleaned) and foreign debris. If the medium contains an excessive

Figure 3.5

Components of the Direct-Fired Rotary Desorber System

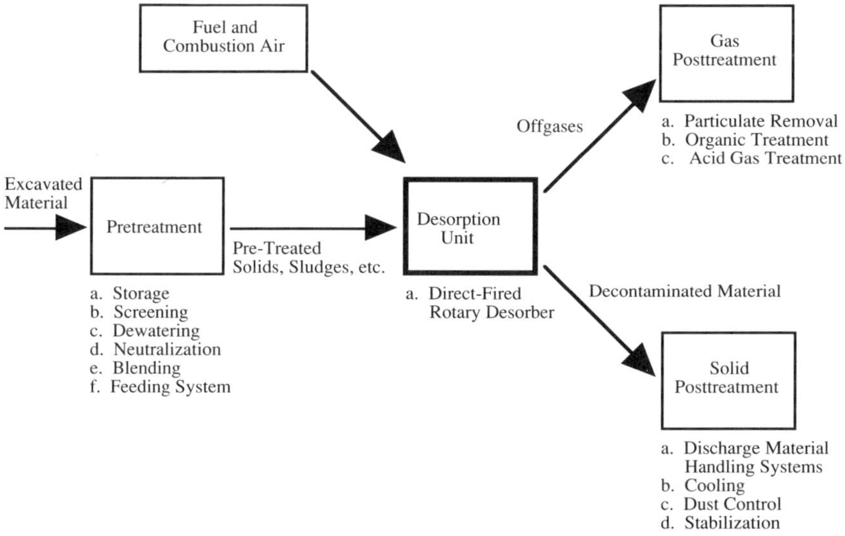

amount of contaminant or moisture, it is usually blended to effect uniform contaminant and moisture levels. Blending can be quite difficult. Excessively wet media can be dewatered by such methods as filter presses and adsorbent addition to 30 to 50% by weight moisture, improving material handling and reducing energy costs. In addition, dry treated solids are sometimes added to effect uniform moisture conditions. In northern climates, the medium may be temporarily stored in a building to preclude freezing. Highly acidic media may also be pretreated (i.e., neutralized) to mitigate corrosion of materials handling, thermal desorption, and treated solids handling equipment.

3.4.1.2 Desorber

The function of the desorption unit, the rotary desorber, is to heat the medium to a sufficient temperature and maintain it for a sufficient period to desorb the moisture and the contaminants from the medium. A rotary desorber consists of a rotating cylindrical metal drum that is inclined slightly with respect to the horizontal axis. Material is passed through the rotating cylinder and is heated by direct exchange with a support flame and/or combustion products. The burner is usually fired with natural gas, propane, or fuel oil. The direction of solids flow through the unit can be either cocurrent or countercurrent with respect to the gas flow direction. Typically, lifters are attached to the inside surface of the cylinder to enhance gas/solid contact; hence, heat and mass-transfer limitations are reduced within the unit. The residence time of the material in the desorber is controlled by cylinder length/diameter ratio, rotation rate, the angle of inclination, and the design of the lifters.

3.4.1.3 Posttreatment

Posttreatment entails the addition of water to cool the treated media and to control fugitive dust emissions. Mixing is usually accomplished in a pug mill or similar unit. Depending upon the nature of the original contamination, the treated material may be redeposited on the site or used in landfills. If the media contains high levels of heavy metals, a stabilizer, such as lime, can be added to the soil in the pug mill. Stabilization processes are discussed in detail in another monograph in this series.[1]

---

1 . See *Innovative Site Remediation Technology: Stabilization/Solidification* — Ed.

The objective of the gas posttreatment system is to remove pollutants from the purge gas stream before it is discharged to the atmosphere. These pollutants consist of the original contaminants and any by-products. Special measures may be needed to handle heavy metals, sulfur dioxide (SO_2), oxides of nitrogen (NO_x), hydrochloric acid (HCl), and other acids depending on waste, fuel fired, and emission requirements. In some cases, National Ambient Air Quality Standards (NAAQS) will need to be considered. Entrained particles must also be removed before discharge. Typical gas posttreatment equipment includes cyclone separators, a secondary oxidizer (typically an afterburner or catalytic oxidizer), a baghouse (filtration system), a scrubber, and an evaporative cooler. Carbon adsorption filters are less common. Depending upon the application, applicable air quality regulations, and chemical constituents and their concentrations, some or all of these posttreatment components may be used.

3.4.2 Status of Development

The direct-fired rotary desorber technology is based on techniques used in such processes as asphalt and cement production, calcination, and common industrial drying processes. The use of either mobile or stationary systems utilizing rotating drums to process granular materials is well established, and direct-fired rotary desorbers are similar to conventional industrial units. Asphalt aggregate dryers have been directly applied in treating petroleum-contaminated soils; where the properties of the soil were suitable, the treated soil has been incorporated into the asphalt product. Because many direct-fired rotary desorbers are adaptations of existing equipment, there is a general uniformity in design and operation.

The principal application of direct-fired thermal desorption units is the treatment of petroleum-contaminated soils. The contamination has often resulted from leaking underground storage tanks (UST). Most of these petroleum-contaminated soils are exempt from regulations under federal hazardous waste laws. The most common exception is soil contaminated with lead from leaded gasoline. Soil is not exempt in any event if it exhibits toxicity characteristic under RCRA waste codes D004 through D017 (Troxler et al. 1992), and treatment of petroleum-contaminated soils must comply with state regulations. Troxler et al. (1992) and the US EPA (1992g) report that one-half to three-fourths of existing rotary dryers used to remediate petroleum-contaminated soils are of the asphalt aggregate dryer design. These asphalt aggregate dryers are not usually designed with afterburner systems. Because of the requirement to meet regional air quality standards, specifically with regard to potential hydrocarbon emissions, some states are requiring that asphalt kilns treating petroleum-contaminated soils be equipped with an afterburner (US EPA 1992g). Thus, while existing technologies have been directly applied to treat petroleum-contaminated soils, the technology is evolving in response to regulatory concerns.

Direct-fired thermal desorbers have been successfully used also to treat wastes regulated under the Comprehensive Environmental Response, Compensation, and Liability Act (CERCLA), Superfund Amendment Reauthorization Act (SARA), and RCRA. Although the systems employed have some of the characteristics of conventional rotary desorbers, they have been specifically designed to treat material contaminated with hazardous wastes. Because of more stringent regulatory criteria, they must have the capability to destroy hazardous compounds desorbed from the solids. Therefore, systems designed to treat hazardous wastes are typically equipped with a gas posttreatment system, such as, an afterburner, carbon adsorption system, or catalytic oxidizer.

3.4.3 Pretreatment Requirements

The pretreatment processes are those used in storing the excavated media, conditioning the material to meet the feed specifications of the desorber, and delivering the material to the desorber.

3.4.3.1 Storage

The contaminated media may need to be stored after excavation. For example, an operation might involve excavation only during the dayshift. Excavated material is often stockpiled to provide an adequate feed supply for continuous operation of the treatment facility. The material should be stored under a canopy to prevent addition of moisture from rainfall. If the contaminated material is stored in a confined location, consideration must be given fugitive emission control and/or ventilation. Frozen media should be warmed before it is fed to the desorber. The storage area should be designed to control runon and runoff precipitation.

3.4.3.2 Solid Particle Size Distribution

The maximum range of particle size that can be treated in most rotary desorbers is 5 to 6.5 cm (2 to 2.5 in.), primarily because of materials handling limitations. Large particles are either screened and/or crushed before treatment, removed and manually cleaned, or returned to the site, if this is permitted. Magnetic objects are usually removed by a magnet suspended over the belt feeder.

3.4.3.3 Contaminant Characterization

In treating contaminated solids, the type and concentration of contaminants in the feed matrix is a key consideration. The LEL of the combustible material in the desorber must be considered, since oxygen is present in the gas. In good practice, concentrations are limited to 25% of the LEL. In addition, materials handling limitations must be considered for wastes containing heavier, tar-like contaminants. Lower explosive limits can be found in the literature for example, in NFPA Standard 325, Sax 1989; Turner and McCreery 1981; and Lide 1990.

The material is generally not uniformly contaminated. In some cases, material with higher levels of contamination can be blended with other less contaminated material in order to make the feed more uniform. But blending is difficult and a uniform feed can not always be produced.

In order to limit equipment corrosion, highly acidic media can be treated with lime in order to maintain a pH greater than 5.

3.4.3.4 Moisture Content

Moisture affects the amount of energy required to heat the medium as well as the handling characteristics of fine-grained soils. Pretreatment methods include filter presses, air drying, blending with drier material, and mixing with treated fines.

3.4.4 Design Data and Unit Sizing

Direct-fired rotary desorbers are increasingly used because of their flexibility and versatility, enabling them to handle the wide variation in conditions encountered among and within sites.

A site is initially characterized by analyzing many solid core samples to develop maps of the type and concentration of contamination, the matrix, and the moisture levels. The maps are used to determine process equipment requirements. Treatability studies are normally conducted in laboratory-scale equipment, in parallel with the mapping. These studies are described in Subsection 4.1.2.

Based upon information provided by the maps and treatability studies, operating conditions can be determined. Combustion gases provide heat to effect the solid temperature required. Direct-fired thermal desorber heat duties commonly range from 7 to 100 MM Btu/hr (Cudahy and Troxler 1992). As a rule of thumb, a heat input of 25,000 Btu/hr is the maximum required for each cubic foot of internal kiln (desorber) volume. Residence time is varied by adjusting the horizontal inclination and rotational speed of the desorber. Typical rotation speeds range from 0.25 to 10 rev/min. Length-to-diameter ratios vary from 2:1 to 10:1.

Another important design factor is the direction of flow of the combustion gas relative to the flow of the solid in the desorber. The flow configuration of the desorber (cocurrent or countercurrent) will affect the arrangement and size of components used in the gas treatment process. In the cocurrent configuration, the gases exiting the desorber are relatively hot. The larger entrained particles can be removed by a cyclone, but the gases are too hot to enter the baghouse. Therefore, the most common design with the cocurrent configuration places the cyclone after the desorber, followed by an afterburner, a gas cooler, a baghouse, an induction fan, and the stack (see figure 3.6a on page 3.17). Since the afterburner is upstream of the baghouse, the particles collected in the baghouse should be essentially free of contaminants in this configuration. If acid neutralization is required, a scrubber may also be included. Venturi scrubbers followed by carbon adsorption units can be used in place of afterburner/baghouse systems.

Alternatively, if the countercurrent flow configuration is used, the gases leaving the desorber will generally be cool enough to flow directly from the cyclone into the baghouse. Thus, the arrangement of the gas posttreatment equipment with the countercurrent desorber configuration is as follows: desorber, cyclone, baghouse, induced draft fan, afterburner, and stack (see figure 3.6b on page 3.18). Since the afterburner is near the end of the pollution control train, the flow rates through the other components are reduced lower than those of the cocurrent configuration. The lower flow rates permit the use of smaller gas posttreatment equipment.

Because of the cooler temperatures in the baghouse of the countercurrent configuration, it is possible that heavier organics will condense in the baghouse and pose a fire hazard or blind the bags. Therefore, heavier organics are not usually treated with this arrangement. In addition, since the afterburner is located downstream of the baghouse, the particles collected in the baghouse may not yet be fully decontaminated and, therefore, are typically recycled back to the desorber.

Figure 3.6(a)

Cocurrent Desorber — Offgas Posttreatment Process

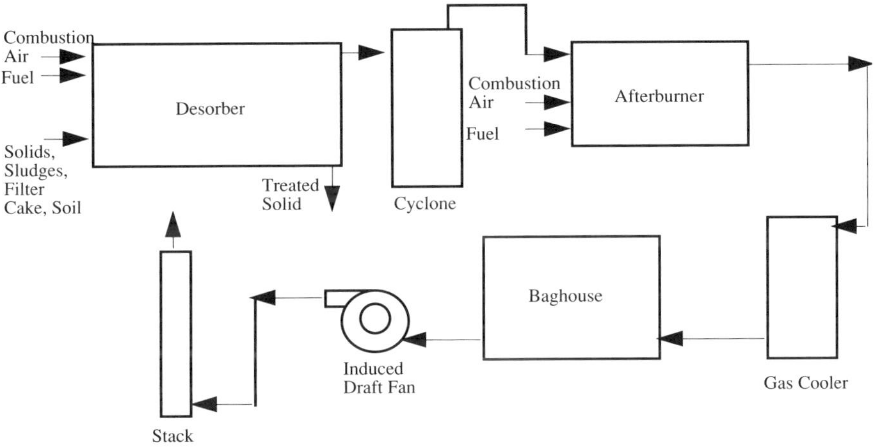

3.4.5 Posttreatment Requirements

Posttreatment of both the treated solids and the gas stream leaving the desorber is required. In addition, posttreatment of liquids is required in systems utilizing a wet scrubber.

3.4.5.1 Solid Posttreatment

Posttreatment of solids typically entails water quenching to cool the solid and control dust. The solid leaves the desorber and usually drops onto a screw conveyor. Water may be added in either the screw conveyor or pug mill. Other

Figure 3.6(b)

Countercurrent Desorber — Offgas Posttreatment Process

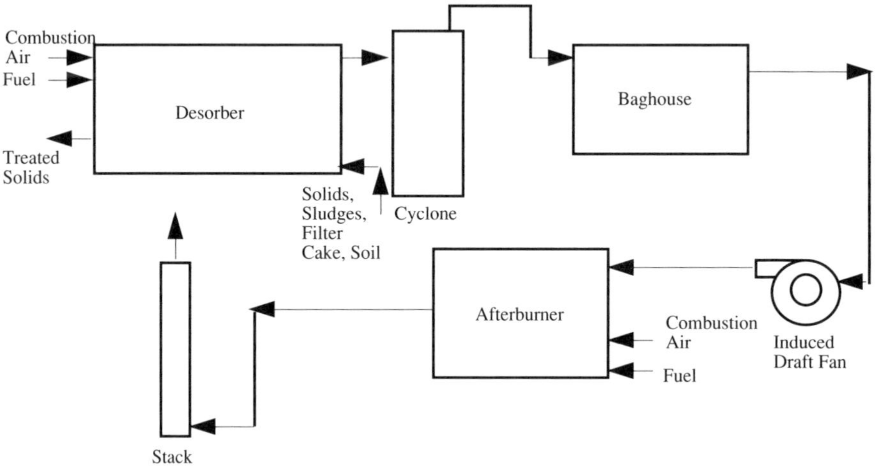

treatments, such as stabilization, may also be necessary if heavy metals are present. Where water, nitrogen, phosphorus, organic material, or other nutrient is absent from the treated matrix, it must be added to solids that will be used as a final cover in order for them to sustain viable plant growth.

3.4.5.2 Gas Posttreatment

The gas posttreatment system removes pollutants from the gas stream before it is discharged. These gases may consist of the original contaminants, the combustion gas products, products of incomplete combustion (PICs), and particulate matter. The gas posttreatment system for a direct-fired rotary desorber requires a larger capacity air pollution control (APC) train than those of indirectly heated thermal desorption treatment systems because of increased gas volume. Equipment used in the gas posttreatment process includes cyclone separators, secondary oxidizers (an afterburner or catalytic oxidizer), baghouses (filtration systems), scrubbers, evaporative coolers, carbon adsorption filters, and condensers. Some of these components are described below.

Organics Control. In systems requiring organics control, a conventional afterburner is the most commonly used oxidizer. Catalytic oxidizers have been used for secondary oxidation to a lesser extent. Carbon adsorption systems have also been used in place of an oxidizer.

The gas treatment train may or may not include an afterburner, depending on the particular application and the regulatory status of the effluents. Most asphalt dryers currently processing petroleum-contaminated soil are not equipped with afterburners (US EPA 1992g) although, unsatisfactory levels of organics were emitted without gas treatment. The likelihood is increasing that gas post-treatment will be required in rotary desorbers processing all kinds of wastes. Conventional afterburners are currently the equipment of choice primarily because of their robustness and relative low cost.

Typically, the afterburner is a refractory-lined shell providing enough residence time at a sufficiently high temperature to destroy organic compounds that have been desorbed from the media. Afterburners typically operate between 760 to 980°C (1,400 to 1,800°F), with a 0.5 to 2.0 second gas-phase residence time. Afterburners can be used before or after particulate control.

Several contractors currently make use of a catalytic oxidizer in place of an afterburner. The catalysts normally used are noble metal compounds, such as platinum or rhodium, used in small quantities and deposited on a support material, such as alumina. In addition to controlling organics, catalytic oxidation can be very effective in eliminating odors.

According to Tarmac Equipment Company, one disadvantage of catalytic oxidizers is that a high-moisture content will adversely affect the operation of a catalytic bed. Furthermore, chlorine and sulfur compounds may poison the bed, resulting in inefficient conversion. The catalytic oxidizer, therefore, must be located downstream from the particulate control and acid gas removal systems. Gases entering the catalytic oxidizer should not have concentrations of organics greater than 25% of the LEL. Catalytic oxidization is addressed in another monograph in the series.[2]

Carbon adsorption has been used to remove low concentrations of organic compounds from the gas phase. The temperature of the process gas should be less than 60°C (140°F). Above this temperature, efficiency may fall. The spent carbon must be periodically regenerated or disposed of. Carbon collection

2 . See *Innovative Site Remediation Technology: Thermal Destruction* — Ed.

efficiency varies with the chemicals of the gases, and selection of carbon adsorption, therefore, depends on the contaminants in the media.

Two important design parameters for carbon adsorption units are the empty bed contact time and superficial gas velocity. The empty bed contact time is the ratio of empty bed volume to the volumetric gas-flow rate through the bed. The superficial gas velocity (or empty bed velocity) is the ratio of the volumetric gas-flow rate to the cross-sectional area of the bed. These parameters are used in estimating the operating period before breakthrough. One report (PRC Environmental Management, Inc., Versar, Inc., and Radian Corporation 1991) suggests a typical empty bed contact time of 4.2 seconds and a superficial gas velocity of 0.3 m/sec (0.9 ft/sec). In reports of other satisfactory applications of carbon absorption used as a tail gas scrubber, the contact time is as low as 2 sec and superficial gas velocities are as high as up to 0.46 m/sec (1.5 ft/sec). Before a specific application is undertaken, however, a trial test to acquire engineering data is recommended.

Removal of Acid Gas and Particles. In addition to organics, acid gases, such as, hydrogen chloride and sulfur dioxide, may also have to be removed, depending on the waste, fuel fired, and system. Particulate removal is almost always necessary.

Conventional venturi scrubbers have been used to remove sulfur dioxide and hydrogen chloride. An added benefit of the venturi scrubber is its capability to remove particles larger than 5 μg in the gas stream (Wark and Warner 1981); however, the resultant water stream and/or sludge must be handled. A significant problem associated with the venturi scrubber is the erosive effect of the gas/liquid mixture passing through the throat section, which is heightened by the high turbulence in this section.

The heart of the venturi scrubber is a venturi throat where gases pass through a reduced area reaching velocities in the range of 60 to 180 m/sec (200 to 600 ft/sec), enhancing mixing. Typically 8 to 45 L (2 to 12 gal) of water per 28 standard m^3 (1,000 standard ft^3) of gas is required in the throat section. High efficiency venturi scrubbers have a pressure drop of 10"-30" w.g.

As the high-velocity gas stream removes gases, particles, and droplets from stack gases, a large number of fine water droplets are formed and entrained. Manufacturers usually provide devices to remove the entrained liquid droplets.

Most thermal desorption systems do not produce significant quantities of acid gases. Acid neutralization may be required, however, to prevent corrosive attack on steel and other materials throughout the system, including the stack.

A few systems employ wet scrubbers. The scrubbers are designed to use an alkali reagent to acid gas stoichiometric ratio of slightly over one. Sodium hydroxide is typically used for pH adjustment. Normally, the scrubbers operate within a pH range of 5 to 7. At higher pH levels, insoluble forms of calcium carbonate and sodium bicarbonate can form and may foul scrubber internals. If heat is being recovered from the stack gas, the gas should not be cooled below its dew point. Stacks can be lined with refractory or fiberglass-reinforced plastic (FRP) to prevent corrosion.

The cyclone separator is designed to remove the largest of the entrained particles from the gas stream. Since it is usually located directly downstream from the primary desorber, the particles collected may still have high concentrations of adsorbed organics and may need to be recycled through the system.

There are wet and dry cyclone separators, but only the dry are presently in thermal desorption systems. The dry cyclone separator is a true inertial separator. Particles entrained in the gas stream enter the cyclone, are directed into a vortex flow pattern, collect on the wall of the separator because of inertial effects, and eventually drop to the receiver part of the unit. Wet cyclone separators operate on the same principle, but use water to assist in gas cleanup and particle entrainment.

Cyclone separators are most efficient in removing larger particles (>15 μm). Agglomeration may occur if the dust is fibrous, sticky, or hygroscopic or if the gas stream contains excessive particulate matter.

Collection efficiencies usually increase with increases in inlet velocities, which is limited, however, by the allowable pressure drop of the separator. A typical inlet velocity is approximately 25 m/sec (80 ft/sec) (PRC Environmental Management, Inc., Versar, Inc., and Radian Corporation 1991).

The baghouse contains cloth filters that collect finer entrained particles. Baghouses contain a series of permeable bags that allow the passage of gas but not particulate matter. Depending on the location of the baghouse relative to the afterburner, the particles collected in the baghouse may be contaminated with organic compounds.

Baghouses are normally used to remove particles <10 μm and are highly efficient in removing particles <1 μm. There are a number of design factors to consider when selecting a baghouse, including the degree of filtration, bag life, ability to clean the bags, ability to provide adequate gas and dust distribution, and dust removal. Typical filter fabrics and suggested operating exposure temperatures are as follows (Bruner 1985):

- nomex (220 to 260°C) (425 to 500°F);
- fiberglass (290 to 315°C) (550 to 600°F); and
- Teflon (230 to 260°C) (450 to 500°F).

The collected particles must be removed from the bags periodically to avoid high-pressure drops. A number of methods have been developed to discharge collected particulate matter on a regular basis, including shakers, compressed air jets, sonic cleaners, and reverse air flow.

3.4.5.3 Liquids Posttreatment

For systems utilizing wet scrubbers, blowdown must be filtered or treated before release. Granular filters are used to reduce total suspended solids. Liquid-phase carbon is used to remove organics from the blowdown. If the blowdown is to be released through a National Pollutant Discharge Elimination System (NPDES) or publicly owned treatment work (POTW), other cleanup parameters may apply.

3.4.6 Special Health and Safety Considerations

In pretreatment, potential safety concerns include explosion hazards in improperly ventilated storage buildings and exposure to fugitive emissions. In the desorber, which is usually operated with excess air, the concentration of contaminants in the feed must be low enough so that 25% of the LEL is not exceeded. Further, in order to control fugitive emissions, most direct-fired rotary desorbers are operated under slight negative pressure. In gas posttreatment, a potential fire hazard exists in the baghouse if hydrocarbons or other combustible materials are allowed to collect on the filters. This presents a potential problem especially in the countercurrent rotary desorber configuration when used to treat material contaminated with heavier organics. The usual precautions relating to hot operating equipment, such as, warning signs, barriers, and safety shields, must be implemented.

3.4.7 Operational Requirements and Considerations

3.4.7.1 Temperature Requirements and Limitations

The temperature required to heat the solids to a temperature sufficiently high to evaporate contaminants from the media depends on the contaminant vapor pressure, initial contaminant concentration, intended cleanup level, and the material matrix. Contaminated materials will vary, exhibiting different desorption characteristics at each site. Laboratory-scale and/or pilot-scale treatability studies addressing the specific solid/contaminant/moisture matrix are usually needed to determine the temperature required.

Controlling the solid temperature is very important in most desorption processes. The temperature is usually directly measured by a thermocouple imbedded at the discharge end of the rotary desorber. Where it is not possible to measure the temperature directly, the retention time and the exiting gas temperature are used as alternative indicators.

The materials of construction chosen for the desorber determine the maximum temperature to which the solid can be heated. Most rotary desorbers are constructed from metal cylinders. According to Troxler et al. (1992), the maximum solids' operating temperature of a desorber made of carbon steel, is 315 to 340°C (600 to 650°F), while those made of alloy steels are up to 650°C (1,200°F).

3.4.7.2 Solid Residence Time

The complexities of contaminated solids require that laboratory and pilot-scale treatability studies be conducted to determine not only the required temperature, but also the length of time for which the solid must be maintained at this temperature.

Troxler and coworkers (1992) report that residence times of petroleum-contaminated soils, in directly-fired thermal desorption devices are usually less than 10 minutes. The residence time of semivolatile organic compounds (SVOCs), can be as long as 20-30 minutes.

The solid residence time may be controlled by adjusting the rotation rate and angle of inclination of the desorber and varying feed rate, although the angle of inclination is usually fixed by the vendor. Residence time of the solid material in minutes is:

$$t = \frac{0.19L_T}{(rpm)(D)(S)}$$

where:

L_T is length of the kiln in meters

rpm is revolutions per minute

D is the ID in meters

S is the slope in m/m

3.4.7.3 Solid Particle Size Distribution

When clay and silt-type soils are treated, "dusting" may increase; that is, particles may become entrained in the gas phase and be carried over to the gas-treating equipment. This is of particular concern because these particles may not be fully decontaminated. Entrainment of particles will also increase the pressure drop through the gas-treatment system which includes a baghouse, and thus increase the power required by the induced draft fan.

The characteristics of the solid can affect also the amount of contaminant that is adsorbed on the solid. For example, smaller particles have greater surface area on which contaminants can adsorb. In addition, some organic compounds adsorb preferentially to solids with high organic contents.

When soil is being decontaminated as part of an asphalt production process, the particle size distribution is very important, because this distribution greatly affects the characteristics of the asphalt. In general, silts and clays are not suitable for mixing with asphalt. The high surface area of these particles degrades the quality of the asphalt (Troxler et al. 1992). Usually materials in the asphalt product must be less than 6% by weight of 74 μm (200 mesh). Soils with a high organic content (peat) are also unsuitable for asphalt (US EPA 1992f).

3.4.7.4 Contaminant Characterization

Since direct-fired thermal desorbers are operated with excess air, the concentration of contaminants must be sufficiently low to the point that 25% of the LEL in the unit is not exceeded. Combustible gas monitors are recommended for monitoring the desorber gases. Blending of solids may be required to stay below the LEL.

The type of contaminant may dictate the selection of the gas posttreatment system. For highly regulated wastes, a more detailed characterization of contaminants and evaluation of the potential for formation of PICs in the thermal desorption process may be required.

3.4.7.5 Moisture Content

Energy required to evaporate moisture in the media increases greatly directly with the amount of moisture. In addition, high moisture levels may cause operating difficulties. For example, moisture may cause the soil's plasticity limit to be exceeded, which may in turn cause the soil to stick to surfaces of the dryer (US EPA 1992g). Moisture may also cause fine particles to form larger clumps with low surface area-to-volume ratios, making the material more difficult to heat (Troxler et al. 1992).

3.4.7.6 Gas Flow Rate

Direct-fired thermal desorbers produce the largest volume of offgas per ton of material of any of the thermal desorbers. This is due to the presence of combustion products from the field used to provide heat for the process. Excessive flow rates should be avoided in order to allow for the use of smaller APC equipment and to minimize dust problems.

3.4.7.7 Desorber Rotational Speed

Solids residence time and the degree of mixing effected are directly related to the rotational speed of the rotary drum. Both of these may affect heat and mass transfer within the desorber. Increased rotational speed may increase particulate entrainment.

3.4.8 Process Variations per Vendors

Since direct-fired rotary desorbers are based upon existing rotary dryer technology, many units on the market have common characteristics. Although some systems have unique features, all have the main components discussed here: pretreatment system, desorption system, and posttreatment systems for both the solid and the gases. Uniformity in design is particularly evident in the asphalt aggregate dryers used to treat petroleum-contaminated soils.

As explained in Subsection 3.4.7, however, variations in design are needed to accommodate different solid/contaminant matrices and to meet varying regulatory requirements. Many vendors offer a variety of designs to suit specific applications. The common design variations and their applications are summarized below.

3.4.8.1 Stationary vs. Mobile Units

Both stationary and mobile units are currently in use. A number of both mobile and stationary facilities are available for treating petroleum-contaminated soils; however, all thermal desorption systems that are currently treating CERCLA wastes are mobile.

3.4.8.2 Secondary Oxidizer

Many asphalt plant aggregate dryers converted to treat petroleum-contaminated soils do not have afterburners or other oxidizer systems. Although this apparently has been common practice in treating petroleum-contaminated soils, some state regulatory agencies are beginning to disallow this practice (US EPA 1992g). Alternatives to conventional afterburners include catalytic oxidizers and activated carbon adsorption units. Use of these alternatives has been limited.

The location of the secondary oxidizer where used, typically depends on whether the desorber is operating in a cocurrent or countercurrent mode. For cocurrent operation, the secondary oxidizer is usually located upstream from the baghouse. This arrangement allows the treatment of heavier organic compounds. For countercurrent operation, the secondary oxidizer is located downstream from the baghouse, and heavier hydrocarbons are not usually treated with this arrangement.

3.4.8.3 Use of Carbon Adsorption

Carbon adsorption filters have been used to treat the offgases prior to discharge. This design does not require an afterburner or other secondary oxidizer. The adsorption bed must be periodically regenerated. If the carbon becomes contaminated with PCBs, it may have to be sent off site for incineration at a Toxic Substance Control Act (TSCA)-permitted facility.

3.5 Rotary Desorber — Indirect Fired

3.5.1 Description

The indirect-fired rotary desorber consists of three subsystems, discussed in Subsection 3.4.1: pretreatment and material handling, the desorption unit, and posttreatment, including offgas treatment and treated solid handling. Figure 3.7 illustrates a typical indirect-fired rotary system with possible variations.

3.5.1.1 Pretreatment

The pretreatment considerations are basically the same as those for direct-fired desorbers, discussed in Subsection 3.4.1.1.

Figure 3.7

Components of the Indirect-Fired Rotary Desorber System

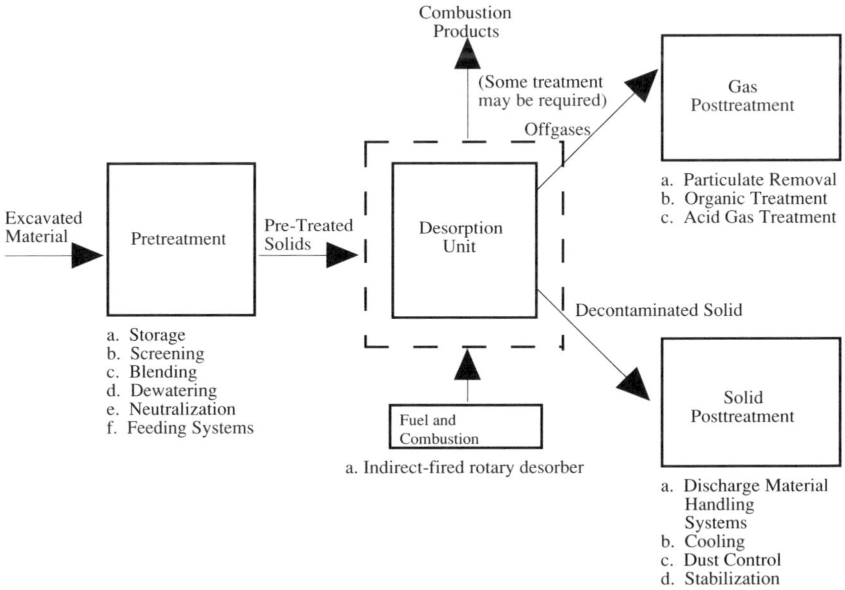

3.5.1.2 Desorber

The desorber unit of indirect-fired rotary systems is different from direct-heated rotary desorbers in one major way — the combustion gases do not come in contact with the solid media. The metal rotary shell is heated on the outside by the combustion of natural gas or propane. The hot shell indirectly heats the solids tumbling on the inside via conduction through the metal shell. As explained by Owens et al. (1991), at higher temperatures radiation may control the heat transfer. Refractory lining is not used because it would impede the heat transfer to the solids and it is not needed for low temperature operation. A sweep gas is used to transfer the volatized organics and water to the offgas treatment system. The thermal desorber is under negative pressure that is induced by a fan downstream of the desorber. Cocurrent flow of the solids and gases is used, normally with lifters inside the desorber enhancing the solid-gas contact. Residence time is controlled by varying rotation speed, the angle of inclination, the lifter design, and the feed rate.

3.5.1.3 Posttreatment

The posttreatment for organics involves one of two general control approaches: destruction or recovery. The destruction processes are discussed in Subsection 3.4.1.3, above; they are not normally used with indirect-fired units. The recovery system uses condensation and refrigeration units, substantially reducing the volume of gases that must be subsequently treated.

Gases of combustion of auxiliary fuel used to heat the desorber shell do not usually require treatment through APC systems; however, depending on the fuel used, NO_x and acid gases may need to be treated. The gases do not require treatment because they do not come in contact with the contaminated medium, as they are isolated from the offgases containing volatilized organics.

3.5.2 Status of Development

Indirect-fired rotary desorbers were developed from existing materials drying techniques. Removal of volatile materials from solids by indirect heating is a well established approach to avoid contamination of the heated material by the combustion gases or avoid problems with explosive gases. Currently, (August 1992), there is one US full-scale design of indirect-heated rotary desorber and a number of European processes. There are also several pilot plant systems available for treatability studies and possible scale up (see Section 5.3). There are considerable amounts of treatability data for both full-scale and pilot-scale systems. The full-scale US unit has been demonstrated at a CERCLA site.

3.5.3 Pretreatment Requirements

The pretreatment steps consist of storing the excavated material, conditioning the material to meet the feed specifications of the rotary dryer, and delivering the media to the desorber. Most of the requirements are similar to those discussed in Subsection 3.4.3 and they are not repeated here. Only differences are addressed.

3.5.3.1 Storage

See Subsection 3.4.3.1.

3.5.3.2 Solid Size Distribution

See Subsection 3.4.3.2.

3.5.3.3 Contaminant Characterization

Acidic wastes must be neutralized before being fed to the system because indirect desorbers are typically made of mild steel. High levels (>10%) of very heavy organics, such as tars and polymeric materials, can interfere with the materials handling systems by plugging and sticking to surfaces.

3.5.3.4 Moisture Content

Typically, less than 40% moisture is desired, 20% is considered ideal, and 5% is too low because of prehandling dusting.

3.5.4 Design Data and Unit Sizing

There are no large variations in the primary unit design of the two full-scale indirectly-heated rotary desorber systems studied (the US system and one European system). The rotary desorbers are less than ≈2.4 m (8.0 ft) in diameter and have heated lengths less than 14 m (45 ft). The systems presently in use are cocurrent flow units wherein solids and inert material flow in the same direction. Solids retention time is determined by the desorber volume, rotation speed, angle of inclination, and lifter design. Thirty to 120 minutes is a typical range of retention times. Rotation speeds can be as high as 2.5 rev/min. Angle of inclination can vary from 1° to 2° downward, moving the solids toward the exit end of the desorber.

Feed rates vary depending on the waste characteristics and the contaminant residual levels required. Nominal feed rates for the process vary from 1.3 kg/sec (5 ton/hr) to 2 kg/sec (8 ton/hr). Where moisture is high, the feed rate is limited because of heat duty in the desorber.

Propane or natural gas is used to heat the shell via conduction. Full-scale systems heat the shell using zones of independently fired burners to control the rate of volatilization. Energy studies performed on soil containing 14% moisture found that 60% of the total heat fired was exhausted to the atmosphere and the remaining 40% was transferred through the shell to heat the solid material. Of the heat transferred through the shell, 60% was used in evaporating water, 0.8% in volatilizing organics and almost 40% in heating the solids (Lehmann 1991).

3.5.4.1 Gas Flows

Full-scale systems operate with the rotary desorber under negative pressure to ensure no leakage or fugitive emissions. The European system has a high gas-flow rate (6,000 to 10,000 normal m^3/hr (200,000 to 350,000 normal ft^3/hr)) because the offgases flow to a destructive, that is, afterburner APC device (Schneider and Beckstrom 1990). The US system has a very low flow rate because a recovery APC system is used. In this system, a nitrogen blanket reduces the necessity of maintaining the organic vapor concentration below the LEL by keeping the oxygen concentration below 4%. Since the majority of the nitrogen is recycled, only 5 to 10% of the carrier gas is vented to the atmosphere, at approximately 34 to 85 m^3/hr (1,200 to 3,000 ft^3/hr).

3.5.5 Posttreatment Requirements

Posttreatment of both the treated solids and the offgases is required. Requirements are similar to those of the direct-fired system addressed in Subsection 3.4.5, above; only differences are discussed here.

3.5.5.1 Solid Posttreatment

See Subsection 3.4.5.1

3.5.5.2 Gas Posttreatment

One important difference between direct-fired systems and indirect-fired systems is the volume of exhaust gases. The indirect-fired system has a much lower volume of exhaust gases. Harwood (1992) states that the differences between a low temperature desorber with no combustion products entering the afterburner (indirect fired) versus a high temperature desorber with combustion gases entering the afterburner is a factor of 3 to 5 times depending on cocurrent or countercurrent flow of the gas and solid.

3.5.5.3 Organics Control

There are two types of organics control — afterburners (destruction) and condenser systems (recovery). There are regulatory advantages, however, in having a recovery system. Regulations under the RCRA generally do not apply to recovery systems, unless they have been specifically cited as the ARAR for the cleanup. Resource Conservation and Recovery Act Regulations usually are the ARAR for destruction systems. They require that the system be proven to effect a 99.99% destruction and removal efficiency (DRE) and to emit less than 100 ppm rolling hour average CO.

Afterburner. See Subsection 3.4.5.2

Condenser. The recovery system uses an eductor scrubber, primary and secondary condensers, and a mist eliminator to recover the organics and water from the nitrogen gas stream. Most of the nitrogen is reheated and recycled through the desorber. About 5 to 10% of the nitrogen is bled to the atmosphere. This purge stream passes through a 2 μm filter and two carbon adsorption beds in series. This system employs a high-energy scrubber, using direct contact with water, to cool the gas to its saturation temperature. Particulates and an estimated 30% of the organics and considerable water are removed from the nitrogen stream by this device. The primary condenser is air cooled and reduces the nitrogen stream temperature to about 5°C (10°F) over ambient temperature, producing the bulk of the liquid condensate. Refrigeration in the second condenser reduces the nitrogen stream temperature to about 4.5°C (40°F).

3.5.5.4 Acid Gas and Particulate Control

Acid gas removal does not present a problem when a recovery system is used because of the low-production levels of the process. When a destruction system is used for organic control, however, depending on the fuel used and on the fuel used for the combustion offgas from the desorber, some consideration of acid gas removal might be in order. Particulate removal in the recovery system is effected by a cyclone and/or scrubber.

3.5.5.5 Liquids Posttreatment

A considerable quantity of liquids can be recovered by the recovery treatment system. For example, water will be recovered at the rate of about 0.3kg/sec (1 ton/hr) from material that is 20% moisture and is fed at 1.3 kg/sec (5 ton/hr). This water must be treated. The condensed organics and water are physically separated. After carbon treatment, the water is used to wet the solids, and the organics are shipped off for incineration or to recycling facilities. Solids in the organic phase can be a problem. Flocculation and filtration can be used to separate the organic from the solid phases. The scrubber blowdown from the recovery system is filtered and the filtered water is returned to the scrubber. The dewatered solids are reprocessed through the desorber.

3.5.6 Special Health and Safety Considerations

Indirect-fired rotary desorbers present no special health and safety considerations. The condenser systems pose the same kind of concern as may any other system that generates a concentrated, hazardous liquid. These materials must be handled as any other hazardous or toxic substance. The unit must be kept rotating at all times to avoid hot spots and resulting equipment failure.

3.5.7 Operational Requirements and Considerations

Many of the operational requirements and considerations appertaining to the indirect-fired rotary desorber are similar to those of the direct-fired rotary desorber, discussed in Subsection 3.4.7; only differences are discussed below.

3.5.7.1 Temperature Requirements and Limitations

See Subsection 3.4.7.1. The mild steel used in the construction of the present systems will limit the operating temperatures of the units to 315°C (600°F).

3.5.7.2 Solid Particle Residence Time

See Subsection 3.4.7.2.

3.5.7.3 Solid Size Distribution

See Subsection 3.4.7.3.

3.5.7.4 Contaminant Characterization

See Subsection 3.4.7.4.

3.5.7.5 Moisture

See Subsection 3.4.7.5. High moisture levels (>40%) require more fuel and larger residual liquid handling systems (recovery approach) and present additional materials handling problems. In addition, processing rates are lowered.

3.5.8 Process Variations per Vendors

The primary variation is in the treatment of the offgases: destruction versus recovery. The condenser/recovery system is usually transportable on about 10 trailers. The destruction system is a stationary facility. These offgas treatment systems are described in Subsection 3.5.5.2.

3.6 *Heated Conveyors — Indirect and Direct*

3.6.1 Description

Conveyors used in thermal desorption applications consist of screw conveyors, paddle or mixing conveyors, and belt conveyors. Direct or indirect heat is applied to the contaminated media while it is transported or moved in a process conveyor. See figures 3.8(a) (on page 3.34) and 3.8(b) (on page 3.35).

In direct-heated conveyor systems, heat is transferred from a source in direct contact with the material being treated. Sources of heat consist of electric resistance heaters imbedded in the conveyor or a source located in the open space above the contaminated media in the conveyor (fuel combustion or radiant heaters). When electric heating is used, offgases generated during processing are greatly reduced.

Figure 3.8(a)

Components of a Direct-Heated Conveyor System

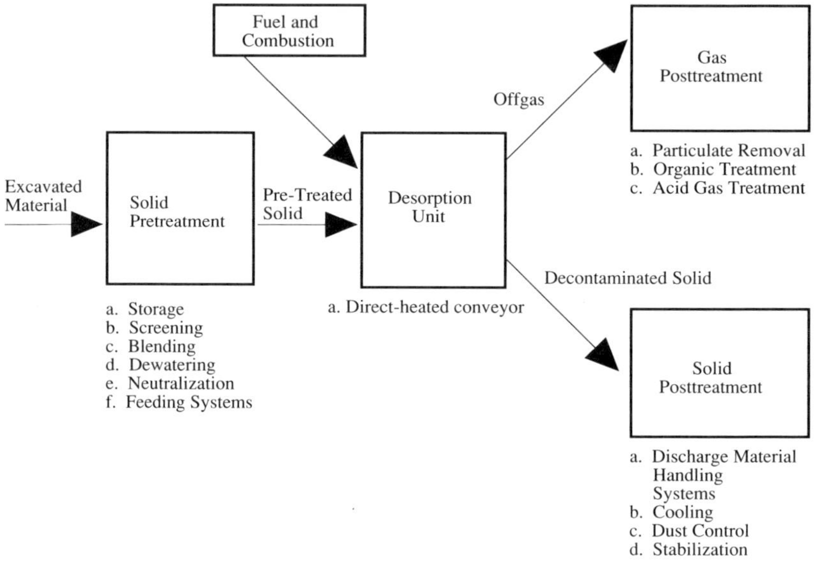

In indirect-heated conveyor systems, the heat is generated outside of the main process desorber in a separate, secondary process unit and is conducted by a media in contact with the desorber conveyor. The source of heat can be the combustion of a common fuel or waste process heat from another process system. Indirect systems employ various media to transfer the heat to the conveyor: steam, special heat transfer fluids (e.g., Dowtherm or Therminol), and eutectic salts. Indirect processes minimize the volume of offgases generated by the thermal desorption system.

3.6.2 Status of Development

Heated conveyors technology is based upon techniques used in mineral processing industries and in bulk solid chemical processing. Many vendors offer services and equipment using heated conveyors. The systems are in various stages of development depending on the specific conveyor and heating

Figure 3.8(b)

Components of an Indirect-Heated Conveyor System

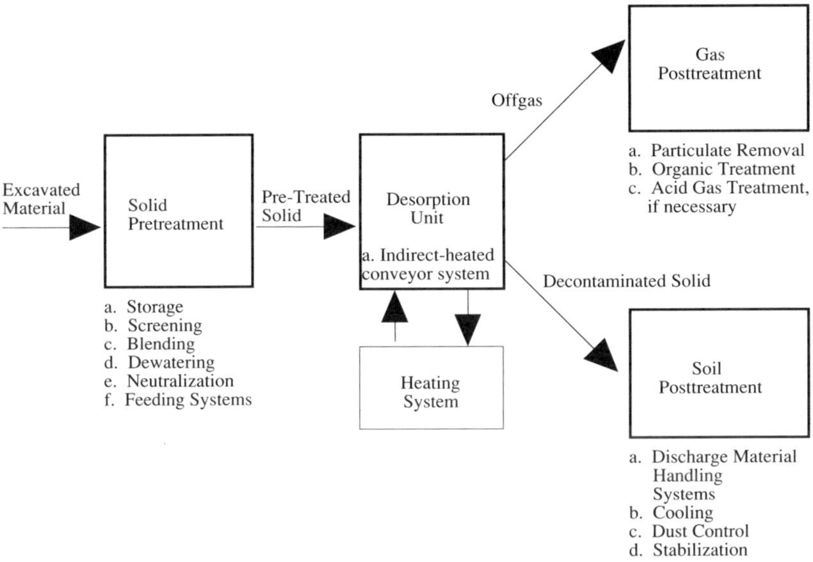

method. Heated conveyors have been used on a full-scale basis to treat petroleum-contaminated soils, sludge from CERCLA sites, and RCRA-regulated hazardous waste. Bench- and pilot-scale units are available to provide demonstration of the full-scale systems. See also table 3.2 (on page 3.36).

3.6.3 Pretreatment Requirements

The pretreatment steps for conveyor desorbers are the storage of the contaminated material, conditioning of the solid to meet the specifications of the desorber conveyor, and delivering the solid to the desorber. Most of the pretreatment requirements are similar to those discussed in Subsection 3.4.3 and are not repeated here. Differences only are addressed.

Table 3.2

Summary of the Status of Heated Conveyor Systems, August, 1992

	Direct-heated system			Indirect-heated system		
	Heat Source			Heating Media		
Conveyor Type	Fuel Combust.	Radiant* Heating	Resistance Heating	Steam	Hot Oil	Salts
Screw	-	Full scale	Full scale	Full scale	Full scale	Full scale
Belt	Full scale	Full scale	-	-	-	-
Paddle	-	-	-	-	Pilot scale	Pilot scale

* Both electric and fuel combustion

3.6.3.1 Storage

See Subsection 3.4.3.1.

3.6.3.2 Solid Particle Size Distribution

In general, it is recommended that particle size be limited to a maximum of 5 to 7.6 cm (2 to 3 in.). But the maximum size of particles that can be treated by a heated conveyor varies with the type of conveyor. For screw conveyors, maximum particle diameters are based upon the screw diameter and size distribution of the particles. See table 3.3 (on page 3.37). Similar limitations appertain to disc or paddle-type conveyors. Belt-type conveyors, are limited to treating the minimum size of particle required to prevent excessive sieving through the belt. Pretreatment steps to reduce particle size include screening, crushing, and shredding.

3.6.3.3 Contaminant Characterization

See Subsection 3.4.3.3

3.6.3.4 Moisture Content

See Subsection 3.4.3.4

Table 3.3

Screw Conveyor Size Versus Maximum Particle Size

Conveyor Size (inches)	Maximum Size (inches) 25% of Lumps
4	0.5
6	0.75
9	1.5
10	1.5
12	2.0
14	2.5
16	3.0
18	3.0
20	3.5
24	4.0

Note: Lumps are materials such as rocks, non-fire material

3.6.4 Design Data and Unit Sizing

The design and sizing of a heated conveyor thermal desorber is keyed to the basic design parameters for treatment systems which have been discussed previously, namely:

- discharge temperature of the solid;
- retention time (maximum of 90 minutes); and
- sweep gas requirements.

The discharge temperature is a critical parameter in the selection and application of a heated conveyor. A conveyor's physical construction should be such that it can operate at the required maximum temperature plus a safety factor for over-temperature excursions. Indirect-heated conveyors should be designed to accommodate the maximum temperature of the conducting media. Typical maximum solid discharge temperatures are (Troxler, et al. 1992):

- Hot oil 150-260°C (300-500°F)
- Molten salt 315-480°C (600-900°F)
- Steam-heated 120-180°C (250-350°F)

The retention time of the conveyor system is determined by the volumetric feed rate of the media and the system's conveying velocity. The retention time of belt conveyor systems, is based upon bed depth because of volatilization limitations and belt speed. Throughput of screw conveyors can be varied with rotational speed, diameter, and flight pitch.

Direct-heated systems generate a greater volume of sweep gas than do indirect-heated systems. By maintaining low-sweep gas velocities, high dust generation and entrainment of particulate from the conveyor can be avoided.

Sizing of a heated conveyor is dependent upon heat transfer calculations, factoring in the overall heat-transfer coefficient from the heat source to the media, which are used to determine discharge temperature. Balancing of operating temperatures, retention times, and conveyor size depends upon the type of conveyor and the heating method. The selection of these parameters is left to the treatment services supplier.

3.6.5 Posttreatment Requirements

Posttreatment of both the treated solid and the offgases is required. Similarities exist between this system and the systems discussed previously. Differences only are discussed below. Where direct-heated systems are used, the gases will include the products of combustion of the fuel source. Indirect-heated systems and electrically heated conveyor systems are sometimes preferred, since they generate a smaller quantity of offgas requiring posttreatment than do other direct-heated systems.

3.6.5.1 Solid Posttreatment

See Subsection 3.4.5.1

3.6.5.2 Gas Posttreatment

The control of emissions of organics from heated conveyor systems will typically be through either thermal destruction in an afterburner or collection of the organics by condensation followed by activated carbon treatment. These methods are discussed in Subsection 3.4.5.2. Particulate removal and liquid posttreatment are discussed in Subsections 3.4.5.2 and 3.4.5.3.

3.6.6 Special Health and Safety Considerations

The heated conveyor systems must be designed to minimize fugitive emissions, which can contain elevated concentrations of toxic gases and chemicals. Conveyors must have adequate safety mechanisms to prevent inadvertent operation during maintenance.

Special attention to containment of the heating media in indirect-heated systems is required to avoid fires and personnel injury. Adequate control of pressure is required where heated media, such as steam and other heat-transfer fluids, are circulated at elevated pressure. Where electrically heated fluids are employed, special attention to high voltage electrical safety precautions is required.

3.6.7 Operational Requirements and Considerations

Operational requirements and considerations are similar to those of direct- and indirect-fired rotary desorbers, discussed in Subsections 3.4.7 and 3.5.7. Systems using electrical resistance heaters have special hardware requirements, such as silicon controlled rectifiers (SCR). The maximum solid temperature will be dependent on the heat transfer media used in the system.

3.6.8 Process Variations per Vendors

A number of units are presently available as explained in Subsection 3.6.2. The conveyor types include belt, screw, or paddle. In addition, a number of offgas treatment systems are available.

3.7 SoilTech System[3]

3.7.1 Description of Process

The SoilTech Anaerobic Thermal Processor (ATP) heats and mixes contaminated soils, sludges, and liquids in a special, indirect-heated rotary dryer.

3. See the monograph in this series *Innovative Site Remediation Technology: Chemical Treatment* wherein this process is addressed as a chemical treatment process — Ed.

The unit desorbs, collects, and recondenses hydrocarbons from solids. See figure 3.9. The unit can also be used in conjunction with a dehalogenation system to destroy halogenated hydrocarbons through a combined thermal/ chemical process.

Figure 3.9

Schematic of the SoilTech ATP Process Illustrating The Four Zone Heating Approach

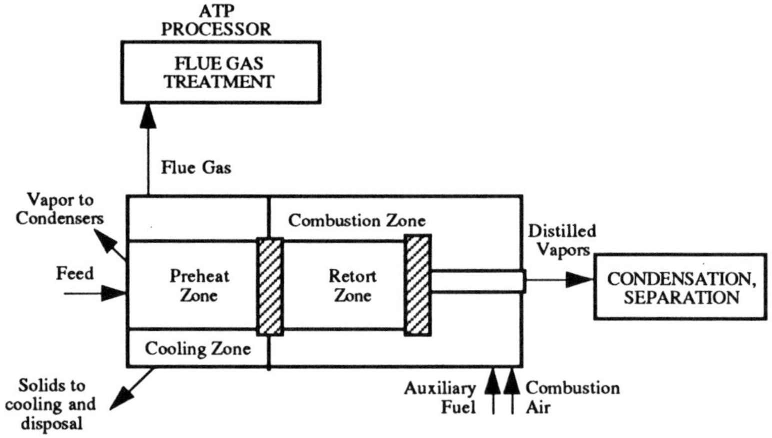

The dryer portion of the system contains four separate internal thermal zones: preheat, retort, combustion, and cooling. In the preheat zone, water and volatile organic compounds (VOCs) are vaporized (temperature, 260°C (500°F)). The vaporized contaminants and water are removed by vacuum to a preheat vapor cooling system consisting of a cyclone to remove solids and a heat exchanger and separator to condense liquids and separate the condensate from the noncondensable gases. Condensed water is usually treated on site, and organics usually require off-site treatment.

From the preheat zone, the hot granular solids and unvaporized contaminants pass through a sand seal to the retort zone (temperature, 510-620°C (950-

1,150°F)). Heavy oils vaporize in the retort zone, and thermal cracking of hydrocarbons forms coke and low molecular weight organics. The vaporized contaminants are removed by vacuum to a retort gas handling system. After cyclones remove dust from the gases, the gases are cooled, and condensed oil is separated into its various fractions. Organics and water are treated off site and on site respectively. The coked solid passes through a second sand seal to the combustion zone, where the coke is burned off the solid and is then either recycled to the retort zone or sent to the cooling zone.

Flue gases from the combustion zone are extensively treated prior to discharge. Treatment is by (1) cyclone and baghouse for particle removal, (2) wet scrubber for removal of acid gases, and (3) carbon adsorption bed for removal of trace compounds.

The treated solid that enters the cooling zone is cooled in the annular space between the outside of the preheat and retort zones and the outer shell of the kiln. Here, the heat from the solid is transferred to the solid in the preheat and retort zones. The cooled treated solid is quenched with water and then transported by conveyor to a storage area.

3.7.2 Status of Development

SoilTech's ATP is a transportable full-scale 3 kg/sec (10 ton/hr) system that has been used twice to successfully remediate Superfund sites with PCB-contaminated soils and sediments. A transportable, pilot-scale system capable of treating 1.5 kg/sec (5 ton/hr) is available as well as bench-scale treatability equipment. Design is complete for a 6.5 kg/sec (25 ton/hr) unit, as of early 1993.

3.7.3 Pretreatment Requirements

Feed with less than 20% moisture is ideal for the SoilTech process; however, less than 5% moisture may cause excessive entrained dust.

Screening to a particle size of less than 5.0 cm (2 in.) is required, and if the solid is mostly fine grained (clay), some sand must be added to the waste to maintain seal integrity with the unit.

3.7.4 Design Data and Unit Sizing

The full-scale SoilTech system, with a nominal rating of 2.5 kg/sec (10 ton/hr), can operate up to a 590°C (1,100°F) solid discharge temperature. The pilot-scale SoilTech system can operate up to 1.5 kg/sec (5 ton/hr) depending on the solid type and sizing. Both systems are transportable on a series of trailers and skids.

3.7.5 Posttreatment Requirements

The residuals generated by the systems consist of the treated solids, condensed water and organics, and offgases. The treated solids, of course, are routinely tested, since it is the object of the remediation. All streams must be handled as described in Subsection 3.4.5.

3.7.6 Special Health and Safety Considerations

In addition to the considerations already discussed, the seals between the preheat and retort zones in the SoilTech system need to be maintained to prevent oxygen from entering the system and creating an explosion potential.

3.7.7 Operational Requirements and Considerations

After several successful site remediations, the SoilTech system is considered mechanically reliable and proven effective. Burning of the noncondensable organics in the combustion zone of the SoilTech ATP may be considered incineration of hazardous waste. The US EPA Regions II and V and state regulatory agencies (New York and Illinois) permitted operation of the SoilTech system as a thermal desorber, as of 1992.

3.8 Environmental Impacts

Air emissions from the excavation, handling, and operation of the thermal desorption system have the potential of impacting the air. The typical on-site remediation, however, is of short duration (less than one year) and thermal desorption systems have extensive air pollution control systems. Therefore, the National Ambient Air Quality Standard (NAAQS) should not be significantly exceeded. The suggestions and approaches discussed in this section can mitigate the short-term, local air impacts.

Excavation of contaminated media for thermal desorption necessarily exposes new material to the atmosphere. Highly volatile contaminants can readily evaporate into the air, presenting potentially dangerous hazards to workers and people living near the contaminated site. A wind screen may need to be constructed around the excavation area to minimize fugitive emissions, or, in more severely contaminated sites, foams, water sprays, organic/inorganic control agents, or portable enclosures may be used to prevent the release of potentially harmful substances. Real-time air monitoring may be used to protect workers at the site and to guard against off-site migration.

It is often necessary to store excavated material, and care must be taken to store it so as to prevent migration of hazardous compounds or objectionable odors from the work area. Physical enclosures with independent dust/vapor control or covers can be used to minimize air impacts.

Equipment used in decontaminating material may develop leaks over time. Every effort must be made to promptly repair leaks and to contain all spills. Wastewater from low temperature thermal desorption units must be treated and tested before being discharged into Publicly-Owned Treatment Plants (POTW) or navigable waterways.

The transfer and handling of residuals can also present difficulties. Treated material exiting the desorber may need to be cooled before transport to its final disposal area. Since the residual matter from the desorption unit must be tested before final disposal, and analyses can take up to several days to complete, treated material may require temporary storage. Dusting can be a problem if the treated material is stored in an open area. Rainwater runoff from ash piles of treated solids can also present problems. Treated material awaiting fixation before final disposal may leach potentially hazardous compounds when it comes into contact with rainwater.

One of the most common complaints from persons residing or working near a remediation site is about the noise. Earth-moving equipment is equipped with warning beepers that sound whenever the machine is backing up. Other noise from large motors or fans may also be bothersome, and noise abatement steps may be necessary.

3.9 *Costs*

Low-temperature thermal desorption of a site contaminated with CERCLA regulated waste is typically conducted by contractors who operate equipment as an on-site service. Contractors provide operating labor and equipment as well as the ancillary services, such as material excavation and waste disposal. Both on- and off-site treatment services are available for treating petroleum-contaminated soils.

3.9.1 Fixed Cost Elements

Fixed costs for operations at a remediation project consist of:

- Planning and Permitting —

  Project team;
  Training;
  Regulatory permit applications;
  Health and safety plans;
  Performance testing;

- Mobilization —

  Site preparation;
  Installation of utilities;
  Transportation to site;
  Excavation, blending, and feed storage pad;
  Containment;
  Storm water control;
  On-site analytical area;
  Land vault; and

- Demobilization —

  Decontamination of equipment;
  Disconnection of utilities;
  Transportation from site; and
  Site restoration.

Planning and preparation costs are those involved in obtaining federal, state, and/or local permits to operate the thermal desorption system. The regulatory environment of projects can vary widely. The cost of applying for air permits can range from a few thousand to many hundreds of thousands of dollars.

Projects being conducted on hazardous waste material must comply with the Occupational Safety and Health Administration (OSHA) regulation 29CFR1910.120, and will require preparation of a proper Health and Safety Plan. Where site-specific permits are needed and hazardous wastes are being treated, emissions testing is often required. One of the author's experience has shown that the cost of emissions testing can range from $15,000, for simple particulate emissions, to over $500,000, for complete stack analysis, which is comparable to incineration trial burn requirements.

Mobilization costs are incurred in transporting the thermal desorption system to the site, assembling it, and connecting utilities. The costs of mobilization depend on the size of the thermal desorption system, condition of the site, and location of the system before mobilization. Thermal desorption systems are highly mobile systems. Many vendors offer systems that can be transported by a few trucks or only one vehicle. Site conditions greatly affect the cost of mobilization. To adequately estimate cost of mobilization, the location of the site and nearest utilities and availability of rights-of way must be ascertained.

The costs of excavation vary with the level of personal protective equipment required (US EPA 1992a). See table 3.4. See US EPA 1992a for several examples of estimated costs for excavation, including Rocky Mountain Arsenal.

If the soil is to be transported to a fixed facility for posttreatment, these costs must also be considered. Estimated costs range from $0.08-$0.15 per ton per mile for petroleum-contaminated soils to $2.00-$4.00 per ton per mile for hazardous wastes (US EPA 1992a).

Table 3.4

Rounded Costs for Excavation Based on Hazard Level

Hazard	Cost (note, $\$1/m^3 = \$1.30/yd^3$)
No Hazard	$22.00 ± $18.80 per m^3
Level D	$75.00 ± $56.00 per m^3
Level C	$90.00 ± $84.00 per m^3
Level B	$120.00 ± $86.00 per m^3
Level A	$130.00 ± $96.00 per m^3

(US EPA 1992a)

3.9.2 Unit Cost Elements

The unit costs consist of variable operating costs, semivariable operating costs, and fixed charges. The variable operating costs include fuel, electric power, process water, and if necessary, neutralization chemicals, nitrogen, and residual disposal (McCormick et al. 1985). The variable costs will depend on the thermal desorption system design and will vary directly with the quantity of material to be treated. They will vary also with the physical and chemical characteristics of the solid. During the thermal desorption process, the inherent matrix moisture will evaporate; the process heat consumption will be a function of the quantity of moisture in the solid. See figure 3.10. Moisture is a critical parameter in an analyses of desorption costs.

Semivariable costs consist of costs of operating labor, administration, maintenance, pre- and posttreatment analyses. They are affected by the thermal desorption system size, regulatory requirements, and local labor costs.

Figure 3.10

Moisture Effect on Heat Requirement Assuming 7.5 ton/hr Facility

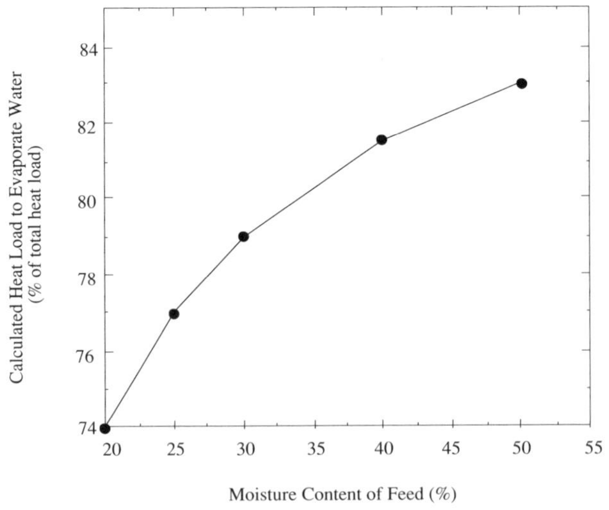

Fixed charges for the operation of a thermal desorption process consist of annualized capitalized cost for equipment and costs of support facilities, insurance, and taxes.

3.9.3 Cost Comparison

To obtain cost information, a search was made of the Vendor Information System for Innovative Treatment Technologies (VISITT) program (US EPA 1992h). The factors that were listed as important in price consideration included moisture (see Subsection 3.9.2), initial concentration and type of contaminant (for example, gasoline, PCBs), contaminant concentration, and the quantity of material to be remediated. The search price range for full-scale systems was \$25-\$1,000 per ton. These values cited in the literature, are quite different from others probably because of the broad range of applications that are considered.

Because of the large variation in cost and little knowledge of the cost effect of moisture, site size, etc., vendors were surveyed for budgetary costs of example thermal desorption projects. The example project sites were divided into two classes, petroleum-contaminated sites and CERCLA or Superfund sites. Both sets of sites had 20% moisture, and the petroleum-contaminanted sites contained 1,000 ppm of total petroleum hydrocarbons contamination, while the Superfund sites had 1,000 ppm of chlorinated VOCs. The vendors were asked to provide costs for 1,000 ton, 10,000 ton, and 100,000 ton sites. The response to the survey was limited and only mobile vendors replied; therefore, the full results are not presented here. The results did show, however, that the size of the site is extremely important. The costs at larger sites are less affected by the initial mobilization and fixed costs, since these costs are distributed over a larger volume of process material. The greater efficiency of larger equipment also helps reduce the unit costs for large sites. In addition, as expected, remediation of petroleum-contaminated sites is less costly than remediation of the Superfund sites.

The four vendors who replied offer different types of thermal desorption equipment. In general, fixed costs were broken down for the Superfund site as follows:

- Mobilization 35-50%
- Demobilization 20-35%

- Plans and Permits (dependent on site size) 15-40%
- Other 0-10%

The unit costs were (% of unit cost, $/ton):

- Fuel 2-20%
- Other utilities 2-10%
- Residual disposal 1-10%
- Operating labor (decreasing with increasing size) 10-50%
- Supervision laboratory 3-10%
- Pre- and posttreatment analyses 5-10%
- Capital equipment 15-30%
- Insurance taxes 2-10%
- Other 5-30%

See US EPA 1992g as an excellent and detailed source for cost estimates for petroleum-contaminated soils.

Costs of operating the SoilTech system were provided by Hutton and Shanks (1993). Fixed costs were $500,000, $1.5 million, and $3.5 million for the three units offered by SoilTech, 1.3 kg/sec (5 ton/hr), 2.5 kg/sec (10 ton/hr), and 6.5

Table 3.5

Cost Data From the Literature

Size of Waste Site (Tons)	Application	Petroleum-Contaminated Soil ($/Ton)	Hazardous ($/Ton)
1,000	Mobile/Transportable	$90-130	$300-600
10,000	Mobile/Transportable	$40-70	$200-$300
100,000	Mobile/Transportable	$35-50	$150-200
All	Stationary	$35-75	Not available

Costs represent total turnkey bid prices.
Sources of Data: Cudahy and Troxler 1992; Troxler et al. 1992

kg/sec (25 ton/hr). Respectively, the unit costs were $300, $200, and $100 per ton. Moisture, particle size, hydrocarbon content, material handling characteristics, and chemical characteristics all affected the costs of remediation. The quoted costs did not include the disposal of generated organic liquid, which in this case contained PCBs.

Costs of operating heated conveyers ranged from $100 to $150 per ton not including excavation, permits, and residual treatment (US EPA, 1992h).

See also table 3.5 (on page 3.48). Again, the size of the site and type of containment are quite important. Troxler et al. (1992) includes data for several different types of desorbers remediating a 20% moisture, silty soil contaminated with 0.3% no. 2 fuel oil. See table 3.6 for excerpts of these data; costs include both operating and fixed costs. Excavation costs, remedial investigation (RI), and project management costs are not included.

Cost information is also provided in some of the case studies presented in Chapter 5.0.

Table 3.6

Costs for Petroleum-Contaminated Soil as a Function of Desorber Type and Site Size

Size (tons)	Cost ($/ton)		
	Large Mobile Rotary Drum (40 MM Btu/hr)	Small Mobile Rotary Drum (10 MM Btu/hr)	Mobile Thermal Screw (12 MM Btu/hr)
2,000	110	70	80
6,000	60	60	60
10,000	50	55	55

Source: Troxler et al. 1992

4

POTENTIAL APPLICATIONS

Thermal desorption has been widely used in treating petroleum-contaminated wastes and is finding increasing use in remediation of Comprehensive Environmental Response, Compensation, and Liability Act (CERCLA) sites. The United States Environmental Protection Agency (US EPA 1992e) states that treatment of volatile organic compounds (VOCs) by thermal desorption is ongoing, planned, or completed at 19 sites, embracing 8 polychlorinated biphenyls (PCBs), 5 semivolatile organic compounds (SVOCs), and 3 pesticides. See table 4.1 (on page 4.2) for the status of thermal desorption projects. Chapter 5.0 presents case studies that also demonstrate the applicability of thermal desorption. See also the draft US EPA *Engineering Bulletin: Thermal Desorption Technologies*, appended as Appendix B (US EPA 1993).

4.1 Determining Applicability — Treatability Testing

The highly variable nature of contaminated material often makes it difficult to determine whether thermal desorption will be effective. Treatability testing of contaminated materials is done to establish process viability and establish design and operating parameters for the optimization of the selected technology. Treatability testing should be conducted as early as possible during the remedial investigation feasibility studies (RI/FS) in order to evaluate the technology and provide a sound basis for the Record of Decision (ROD).

Treatability testing must be carefully planned and executed in order to ensure that sufficient data are generated for use in evaluating the performance (US EPA 1992d). It is typically divided into three phases: (1) remedy screening, (2) remedy selection, and (3) remedy design. Details are given in US EPA 1992d.

Table 4.1

Status of Thermal Desorption, April 1992

Region	Project	Status	Key Contaminants
1	Cannon Engineering, Mass.	Completed	VOCs
1	Re-solve, Mass.	Predesign	PCBs
1	McKin, Maine	Completed	VOCs
1	Union Chemical Co., Maine	In design	VOCs
1	Ottati and Goss, N.H.	Completed	VOCs
2	Caldwell Trucking, N.J.	In design	VOCs
2	Metal tech/Aerosystems, N.J.	Design Complete/Installation not begun	VOCs
2	Reich Farms, N.J.	In design	VOCs, SVOCs
2	Waldick Aerospace Devices, N.J.	Design Complete/Installation not begun	VOCs
2	American Thermostat, N.Y.	In design	VOCs
2	Claremont Polychemical, N.Y.	In design	VOCs
2	Fulton Terminals, N.Y.	In design	VOCs
2	Sarney Farm, N.Y.	In design	VOCs, SVOCs
2	Solvent Savers, N.Y.	Predesign	VOCs, PCBs
2	GE Wiring Devices, P.R.	In design	Metals
3	U.S.A. Letterkenny, Pa.	Predesign	VOCs
3	Saunders Supply Co., Va.	Predesign	SVOCs, Metals (As)
4	CIBA-GEIGY (MacIntosh), Ala.	Predesign	Pesticides
4	Aberdeen Pesticide Dumps, N.C.	Predesign	Pesticides
4	Sangamo/Twelve Mile/Harwell, S.C.	Predesign	VOCs, PCBs
4	Wamchem, S.C.	In design	VOCs
4	Arlington Blending & Packaging Co. Tenn.	Predesign	VOCs, SVOCs, Pesticides, Metals
5	Acme Solvent Reclaiming, Ill.	Predesign	VOCs, SVOCs, PCBs
5	Waukegan Harbor, Ill.	Operational	PCBs
5	Anderson Development, Mich.	Operational	Organics
5	Carter Industries, Mich.	Predesign	PCBs
5	Univ. of Minnesota, Minn.	In design	PCBs
8	Martin Marietta, Colo.	Predesign	PCBs(As)

(US EPA 1992e)

4.1.1 Remedy Screening

The purpose of remedy screening is to determine whether thermal desorption will remove the contaminants of interest. Samples used in testing should be representative of site conditions for the range of concentrations of contaminants. If solids are expected to be blended in full-scale operation, blending should be done before testing. If solids will not be blended, and the character of the material changes drastically from location to location or at various depths, multiple test runs should be conducted to assure that samples from areas of both high- and low-contaminant concentration are tested and that the contaminants are reliably removed from the media.

The particle size distribution of the solids being tested should be consistent with the particle size distribution that will be treated on a full scale.

A test is normally conducted in order to make a rough determination of technical feasibility for low-temperature thermal desorption. In the test, care should be taken to minimize the temperature and concentration gradients so that the solid is at a constant temperature and mass transfer of contaminants to the surface is not limited.

Thermocouples must be used so that the true solid temperature can be recorded. Time-at-temperature data should be recorded, and concentrations of the target contaminants should be measured in the solid both before and after the tests. If a number of contaminants are present, the least volatile contaminant should be monitored. Several test runs should be done to assure that the data are reproducible. The primary criterion used to evaluate a test is the degree to which the target compounds have been desorbed from the contaminated media. If the tests indicate that thermal desorption may be effective in cleaning the media, then further testing should be done to define the operating characteristics of the thermal desorption device.

4.1.2 Remedy Selection

The primary objective of the remedy selection phase is to determine which system will best serve the needs of the job at hand. Materials handling constraints or other constraints may dictate which system should be used. If so, one would proceed directly to the remedy design testing phase.

Remedy selection should be systematic. The work done must satisfy both the site owner and the regulatory agencies. A significant amount of attention must be directed toward establishing data quality objectives and implementing appropriate quality assurance/quality control (QA/QC) programs in data collection and analysis.

Systems typically tested in the remedy selection phase of treatability work include the following:

- rotary desorber — direct or indirect;
- conveyor;
- other technologies; and
- offgas treatment system.

Many equipment vendors, as well as independent outside testing companies, have bench- and pilot-scale thermal desorption units that can be used in treatability testing. Once a particular device (e.g., rotary desorber) has been selected, remedy design testing should begin.

4.1.3 Remedy Design

The main objective of the remedy design phase of treatability testing is to obtain all information necessary to ensure the success of the full-scale treatment unit. Target treatment levels must be established for all compounds of interest and all applicable, relevant, and appropriate regulations (ARARs) must be identified for the specific site before testing so that complete criteria for evaluation can be developed.

Criteria for process evaluation should include not only the cleanup standard for the media requiring remediation, but also the requirements of all federal, state, and local regulations that are expected to apply to any liquid, gaseous, or fugitive emissions from the unit.

In general, sufficient testing should be done to develop and confirm a heat and material balance around the unit. Concentrations of all contaminants of regulatory concern should be measured in the feed material. All streams exiting the desorber should be characterized and analyzed for not only the target compounds of interest, but also for potential intermediates and compounds of regulatory concern, such as dioxins, furans, acid gases, carbon monoxide, heavy metals, and polynuclear aromatic hydrocarbons (PAHs).

Particulate matter exiting the unit should be characterized both in terms of loading and particle size distribution. Gases leaving the unit should be analyzed to determine the content and concentration of compounds that must be removed before discharge into the atmosphere. Any condensed liquids from the offgas treatment system should be analyzed for target and other potentially hazardous intermediate compounds.

Treated solid residue should be analyzed for target compounds and should undergo any necessary further physical and chemical testing to ensure that it can be disposed of appropriately. Baghouse dust or other fines from the unit must be similarly analyzed. To aid in the identification of any problems that might arise in the processing of the contaminated media with respect to the quality of the treated material, these streams should be analyzed separately, even if they will be combined in the full-scale unit.

If carbon adsorption is to be used to treat the offgas, a complete mass balance should be done around the carbon adsorption unit. The issue of disposal or regeneration of the carbon should be addressed during the treatability testing phase to preclude problems in the disposal, regeneration, and usage of activated carbon.

4.2 Quality of Residuals

All thermal desorption units create a number of residual streams that must be properly managed. Both destruction and recovery offgas APC systems create residual streams, such as particulate, scrubber water, condensed water, condensed organic liquids, and stack gases, discussed in Subsections 3.4.5, 3.5.5, and 3.6.5. These residuals can be divided into three types: solids, liquids, and gases.

4.2.1 Solid Residuals

Solid residuals include the treated solids, particulate from the scrubbers, baghouse and micron filters, and used carbon from both gas- and aqueous-phase carbon adsorption beds. Treated solids must meet the cleanup requirements set by the regulatory agencies, but the material may also need wetting and stabilization before disposal. Posttreatment use of the material is a consideration in determining the need for further treatment. Chapter 5.0 discusses in detail the soil residual levels of various contaminants that have been effected by thermal desorption.

The chemical constituents of the particulate collected in the scrubber, baghouse, and micron filter will determine the fate of the residual. Residuals may frequently be considered a hazardous waste requiring further treatment or regulated disposal. With proper collection or destruction of contaminants in the offgas, used gas-phase carbon can be regenerated on site. But generally, it is shipped off site for regeneration, as is the aqueous-phase carbon.

4.2.2 Liquid Residuals

Liquid residuals from thermal desorption include the scrubber waters, condensed water, and condensed organics. When the destruction approach to air pollution control (APC) is employed, the scrubber waters will have particulate, neutralized salts, and possibly organic constituents. Particulates should be handled as explained immediately above.

When the recovery approach to APC is employed, the condensed water is separated from the condensed organics by a phase separator. The liquid organics must be shipped to an incinerator or recycling facility. Aqueous liquids from the phase separator usually contain some level of organics (0.01 to 10 ppm) that must be removed. Following treatment, discharge to an on-site water treatment plant or the local municipal waste treatment plant is possible. In most cases, the thermal desorption vendors treat the recovered water and then use it to wet the treated solids.

4.2.3 Gaseous Residuals

Depending on the type of system used, there are a number of possible gas streams that require further treatment, including:

- purge gas ladened with organics; and
- combustion gas that has been used as purge gas (direct-fired units).

The purge gas and combustion gas that has been used as a purge gas are major emission sources requiring employment of extensive APC systems. These systems are discussed in detail in Chapter 3.0. If any organics are stored in tanks, the tanks should be vented to an appropriate APC system.

5
PROCESS EVALUATION

When considering application of a technology, it is instructive to review past experiences. To this end, this Chapter presents case studies of operations of full-scale systems and reports of operations of pilot- and bench-scale systems.

5.1 *Full-Scale Systems*

There is growing literature concerning the operation of full-scale desorbers, both open and through vendors. The detail and completeness of data provided, however, varies considerably among sources. In reviewing the literature, information should be sought in the following areas: site description and specifications, contaminant types, process performance, process by-products, cost, and operational considerations.

See table 5.1 (on page 5.2) listing case studies of direct-fired rotary desorber operations. In addition, the following subsections present case studies of operations of direct-fired and other systems upon sites for which relatively complete information is available.

5.1.1 McKin Site (Gray, Me.) — Direct-Fired Desorber

The McKin site had previously been used as a liquid waste storage treatment and disposal facility. As a result of these operations, soil at this site was contaminated with volatile organic compounds (VOCs) and heavy oils (Bell and Giese 1991; US EPA 1991a, 1992e; Canonie Environmental Services, Corp. 1991). A drinking water aquifer had also become contaminated. The soil was contaminated with trichloroethene (TCE), tetrachloroethene (PCE), and 1,1,1 trichloroethane (TCA). The contaminant concentrations were >3,000 ppm for TCE, >120 ppm for PCE, and >19 ppm for TCA.

Table 5.1

Direct-Fired Thermal Desorber Case Studies
Available Information as of August 1992.

Source/Reference	Site Description	Contaminant Type	Performance	By-Products	Cost	Operational Considerations	Number of Cases
USEPA 1991b Operator — Weston Services, Inc.	No	Yes	Yes	No	No	No	1
USEPA 1991b Operator — Canonie Environmental Services, Corp.	No	Yes	Yes	No	No	No	3
Canonie 1991 Marketing Literature	Yes	Yes	Limited	No	Yes	Limited	4
Bell and Giese 1991 Operator — Canonie Environmental Services, Corp.	Yes	Yes	Yes	No	No	No	4
Soil Purification Inc. 1991 Marketing Literature	No	Yes	Yes	No	No	No	10
USEPA 1992h Operator — Soil Purification Inc.	Limited	Yes	Yes	No	No	No	10
Halliburton NUS Environmental Corp. 1991 Marketing Literature	No	Yes	Yes	No	Yes	No	2
California Dept. of Health Services 1990a Operator — Earth Purification Engr. Inc.	Yes	Yes	Yes	Yes	Fuel	Yes	1
USEPA 1992g Operators — Various	No	Yes	Yes	Yes	No	No	25
USEPA 1992h Operators — Various	No	Yes	Yes	No	No	No	5

Over 8,800 m^3 (11,500 yd^3) of soil was excavated and treated using Canonie Environmental Services Corporation's Low Temperature Thermal Aeration (LTTA) process. The principal components of this process are a feed system, a direct-fired desorber, a pugmill, cyclonic separators, a baghouse, a venturi scrubber, and a carbon filter system. The soil was treated in a continuous operation with a soil retention time of 6 to 8 min. During processing, the soil was heated to approximately 150°C (300°F).

The treated soil contained VOC concentrations of less than 0.04 ppm (0.04 for TCE, 0.02 for PCE, and 0.02 for TCA) and concentrations of polynuclear aromatic hydrocarbons (PAHs) below 10 ppm. The treated soil was solidified and disposed of on site.

During excavation, VOC emissions were controlled so as not to exceed 2 ppmv at the site boundaries. In addition, fugitive dusting was controlled to remain below 150 mg/m^3 at the site boundaries.

The total cost for treatment of this site was $6,500,000.

5.1.2 Ottati and Goss Site (Kingston, N.H.) — Direct-Fired Desorber

The Ottati and Goss site had previously been used as a facility to treat organic solvents (Bell and Giese 1991; US EPA 1991b, 1992e; Canonie Environmental Services Corp. 1991). The soil and groundwater at the site were contaminated with VOCs and other chemicals. Soils contaminated with polychlorinated biphenyl (PCB) levels above 20 ppm were to be incinerated, according to the Record of Decision (ROD) (US EPA 1987). The original contaminant concentrations were PCE, 1,200 ppm; TCA, 470 ppm; and TCE, 460 ppm. Contaminants such as toluene (3,000 ppm) and ethylbenzene (440 ppm) were also found. Approximately 3,400 m^3 (4,500 yd^3) of soil were excavated and treated. The Canonie Environmental Services Corporation LTTA process was used to treat the contaminated soils at this site. During processing, the soils were heated to approximately 150 to 200°C (300 to 400°F).

This treatment process reduced the contaminant concentrations to the following levels: PCE, TCA, and TCE, <0.025 ppm; toluene, 0.11 ppm; and ethylbenzene, 0.025 ppm. The cleanup criteria was 1 ppm (US EPA 1987).

The total cost for the treatment of this site was $1,470,000.

5.1.3 Cannon Bridgewater Site (Bridgewater, Mass.) — Direct-Fired Desorber

The Cannon Bridgewater Site, formerly a chemical waste storage and incineration facility, was remediated using the LTTA system (Bell and Giese 1991; US EPA 1991b, 1991c, 1992e; Canonie Environmental Services Corp. 1991). Between 1974 and 1980, this seven-acre site was used to handle, store, and incinerate chemical wastes. After operations were halted by state regulatory agencies, 590 m^3 (155,000 gal) of sludge and liquid waste stored in tanks and drums were removed from the site.

A remedial investigation was performed between 1982 and 1987 to determine the level of contamination. The soil was found to be contaminated with VOCs, such as, TCE, vinyl chloride, benzene, and toluene. Pretreatment contaminant concentrations were as follows: PCE, 4 ppm; toluene, 78 ppm; xylene, 29 ppm; chlorobenzene, 57 ppm; and total VOCs, 461.3 ppm. The soil moisture content varied from 5 to 25%.

Approximately 8,400 m^3 (11,000 yd^3) of soil were treated in a continuous operation. Pretreatment included screening, mixing, and dewatering. The maximum particle size of the feed was limited to 5 cm (2 in.). The solids feed rate to the kiln was about 10 kg/sec (40 ton/hr). The LTTA process was used to heat the soils to approximately 230 to 260°C (450 to 500°F). This process reduced total VOCs concentrations to less than 0.025 ppm.

Both the residuals from the air pollution control (APC) equipment and the generated wastewater were treated on site and were disposed of off site. No costs were given for this operation.

Approximately 150 m^3 (200 yd^3) of PCB-contaminated soil were also excavated and incinerated off site. No costs were given.

5.1.4 Caltrans Maintenance Station Site (Kingvale, Cal.) — Direct-Fired Desorber

Soil at this site was contaminated with diesel fuel that leaked from an underground fuel tank (California Department of Health Services 1990a). The diesel concentrations ranged from 440 to 5,200 ppm, with approximately 5% moisture content. Approximately 153 m^3 (200 yd^3) of soil were excavated and treated using Earth Purification Engineering Inc.'s Soil Cleanup System, a direct-fired desorber. The California State Department of Health Services assisted in the treatment and monitored it. During the treatment, several tests were made to evaluate the performance of the system. The treatment system consisted of a reciprocating pan feeder, an asphalt recycling rotary desorber, dual cyclones, an exhaust cooler, a baghouse, and an exhaust fan. It should be noted that this unit was not equipped with an organic emissions control device.

During part of the processing period, stack gases were sampled to estimate emissions. The exhaust gas and soil exit temperatures were 425 and 413°C (796 and 775°F) for this period, and the feed rate to the desorber was 1.4 m^3/hr.

(1.8 yd^3/hr.). Under these conditions, the soil contaminant concentration was reduced from an average 1,875 ppm to less than 1 ppm. Stack gas samples revealed, however, a nonmethane VOC concentration of 268 ppmv and a CO concentration of 1,373 ppmv. Destruction and removal efficiency was estimated to be between 71 and 89%. The concentration of total particulates was 4.56 gr/dry standard cubic meter (dscm) (0.1278 gr/dscf). Even when the desorber was fed decontaminated soil, the nonmethane VOC and CO concentrations were 67 and 545 ppmv, respectively.

During these tests, the reciprocating pan feeder did not deliver a continuous feed to the desorber. This resulted in fluctuations in the heat input to the system and offgas generation, and made it difficult to maintain a vacuum in the desorber.

The total cost of the site cleanup was not reported. The following amounts of fuel were consumed during the processing: 4,350 L of propane (1,150 gal), 200 L of gasoline (52 gal), and 280 L of diesel (75 gal).

5.1.5 Coke-Oven Plant Soils — Indirect-Fired Desorber

Deutsche Babcock Anlagen (DBA) has presented some data on the cleanup of a former coke oven site (Schneider and Beckstrom 1990). The system used was a 2.2 m (7.2 ft) in diameter by 21 m (69 ft) indirect-heated rotary desorber. Capacity of the system was 2 kg/sec (8 ton/hr) for a moisture content of 20%. Temperatures range from 500 to 750°C (930 to 1,380°F) in the desorber, with a secondary combustion chamber for gas destruction operating at temperatures of 1,000-1,300°C (1,830-2,400°F). The wall temperature of the desorber during normal operation is 600-650°C (1,100-1,200°F). Following their passage through the secondary chamber, the gases are cooled, limestone is injected for neutralizing acid gases, and the gases then pass through a baghouse filter. The desorber is heated with hot flue gas produced by 18 natural gas burners installed in the heating jacket.

The soils fed contained various PAHs, 17 of which were directly measured. The soil results are shown in table 5.2 (on page 5.6). Data exist for two tests but only one set of results is presented. As shown in the table, concentrations of most PAHs were below 1 ppm with the exception of phenathrene, fluoranthene, benzo(b)fluoranthene, and benzo(e)pyrene. No information was given about organics in the exhaust gas.

The capital cost of the plant was approximately $5.5MM(US), and operating costs were estimated at $65-80 per ton of soil. It was noted that the cost of excavation, prefeed treatment, and backfilling are dependent on the conditions in the area and can vary over a wide range.

Table 5.2

Results of the DBA Pyrolysis Full-Scale Plant
Konigsborn Coke-Oven Plant

Constituents	Feed	Discharge
Naphthalene	161.6 ppm	0.5 ppm
2-methyl-naphthalene	73.8	0.1
1-methyl-naphthalene	42.9	0.1
Dimethyl naphthalene	93.2	0.3
Acenaphthylene	68.2	0.1
Acenaphthene	42.3	0.1
Fluorene	238.0	0.1
Phenanthrene	1055.3	1.4
Anthracene	226.0	0.3
Fluoranthene	688.6	1.3
Pyrene	398.2	0.6
Benzo[a]anthracene	2259.2	0.3
Chrysene	134.6	0.9
Benzo[b]fluoranthene	168.5	5.2
Benzo[k]fluoranthene	81.9	0.3
Benzo[e]pyrene	111.5	1.1
Benzo[a]pyrene	138.1	0.4
Indeno[1,2,3-cd]pyrene	69.5	0.1
Dibenzo[a,h]anthracene	23.2	0.1
Benzo[g,h,i]perylene	60.2	0.1
Total	6134.8 ppm	13.4 ppm

Source: Schneider and Beckstrom 1990

5.1.6 Wide Beach Superfund Site (Buffalo, N.Y.) — SoilTech

The SoilTech anaerobic thermal processor (ATP) system was used in the cleanup of PCBs from the Wide Beach Superfund Site (Vorum and Ritcey 1991; Vorum 1991, 1992). The site was contaminated through the use of oil containing PCBs on the roads of the Wide Beach residential development to reduce dust. As described in Subsection 3.7.1, the SoilTech ATP system is an indirect-heater desorber with zones of heating and cooling. Total solids residence time in the unit is under one hour (30 to 45 min), and application of temperatures up to approximately 510 to 620°C (950 to 1,150°F) are cited. Feed rates were approximately 2 to 2.3 kg/sec (8 to 9 ton/hr). The added effect of chemical dechlorination of PCBs was applied at the Wide Beach site to process 38,000 kg (42,000 ton) of PCB-contaminated soil.

The average feed concentration of PCBs in the soil was approximately 25 ppm (the range was 10 to 5,000 ppm), and the treatment was shown to reduce the residual PCBs to less than 70 ppb. Stack PCB concentrations were less than 25% of the allowable rate (target was 1.5 g/hr (3.33×10^{-5} lb/hr)) and dioxins and furans were within the state target value of 0.2 ng/dscm. Vorum (1991) states that no PCB-bearing residuals needed to be treated off site, with the exception of process wastewater which had PCB concentrations ≈1 ppb.

No costs are given for the project; however, the authors state that a range of \$150-\$300/ton, not including costs of excavation and delivery of feed materials or pickup and disposal of treated solids and oil product, can be expected. Costs depend upon volume of material to be treated, material handling considerations, moisture level, hydrocarbon content and the potential for recycling, acidity, particle-size distribution, water disposal options, and regulatory impacts. Site remediation was completed in June, 1993.

5.1.7 Waukegan Harbor Superfund Site (Waukegan, Ill.)—SoilTech

The SoilTech ATP system was used also to treat 11,500 tonne (12,700 ton) of PCB contaminated sediments using the 2.6 kg/sec (10 ton/hr) ATP transportable system. An excerpt of the report of Hutton and Shanks (1993) on this project is appended as Appendix C, since the information it provides is quite complete.

5.1.8 Anderson Development Site (Adrian, Mich.) — Indirect-Heated Screw Conveyor

A wastewater lagoon at the Anderson Development Company site was contaminated with VOCs, semivolatile organic compounds (SVOCs), and 4,4' methylene bis (2-chloroaniline)(MBOCA $C_{13}H_{12}Cl_2N_2$). A remediation was conducted using the Roy F. Weston, Inc. (WESTON®) Low Temperature Thermal Treatment System, which utilizes an indirect-heated screw desorber. The screw was heated by circulating heat transfer oil. The feed rate was 0.54 kg/sec (2.0 ton/hr). The soil discharges at 270°C (515°F) after approximately 90 minutes of thermal treatment. The desorber offgases were vented through a baghouse dust collector, an air cooled condenser, a refrigerated condenser, and carbon adsorption bed before discharge.

The chemical of primary concern was MBOCA, a semivolatile compound with a low solubility and extremely low vapor pressure. During a Site Demonstration Test treatment of lagoon sludge, MBOCA removal efficiencies greater than 88% for sludge containing from 10 to 800 ppm of MBOCA were achieved. Treated sludge ranged in concentration from 3 to 9.6 mg/kg (US EPA, 1993). During processing, volatile organics were removed to below detection limits (approximately 60 ppb). Semivolatile compounds generally decreased during treatment; however, some compounds did increase as a result of chemical transformation specifically, chrysene and phenol. Polychlorodibenzodioxins (PCDDs) and polychlorodibenzofurans (PCDFs) were formed, but were removed from the exhaust by the vapor-phase treatment system. During operation, by-products included fabric filter dust, condenser water, and activated carbon from vapor and liquid phase treatment systems. The total cost for treatment of the 3,000 ton of lagoon sludge was $1,700,000 (US EPA 1992c).

5.1.9 Gasoline and Diesel Soil — Direct-Heated Conveyor

A full-scale demonstration of a U.S. Waste Thermal Processing direct-fired belt conveyor system was conducted using a synthetic contaminated soil and reported by the California Department of Health Services (1990b). Natural soil was contaminated with diesel fuel and gasoline for the demonstration. The system operates at temperatures from 150-340°C (300-650°F) using eight burners that are located in the primary furnace. A 2.5-cm (1 in.) layer of soil is moved through the furnace and soil throughput is controlled to maintain the desired soil temperature. The throughput ranges from 3 to 6.9 m^3/hr (4 to 9 yd^3/hr). The offgases are vented to an afterburner for control of emissions of organics and to a venturi scrubber for particulate control. The liquid blowdown from the scrubber is used to cool and wet the treated soil.

During testing, the feed soil was spiked to contain 5,000 ppm of diesel or gasoline. The moisture content of the soil varied from 5.2 to 8.9% by weight. During the gasoline test, stack emission analyses showed no dioxins, furans, or PCBs, with the exception of dichlorobiphenyl, present at 0.073 µg/dscm; the report states that this measurement is questionable. Phenanthrene, anthracene, fluoranthene, and pyrene were also detected. Lead and chromium concentrations were 10.6 and 14.4 µg/dscm, respectively. Some cadmium was also detected. The diesel run stack analyses showed no dioxins, furans, or PCBs. The analyses showed 6.6 μg/dscm of naphthalene; 13 μg/dscm of phenathrene; and 0.25 μg/dscm of anthracene. Some carbon disulfide and methylene chloride were also measured, but these data were questionable. Lead, chromium, and cadmium levels were slightly lower than in the gasoline test. Lead emissions for both tests were below the regulatory standard. During the gasoline tests chlorinated volatiles, fuel hydrocarbons, and aromatic hydrocarbons were not detected in the treated soil. Metals were detected, but at concentrations below 20 ppm, except for barium (76 ppm). During the diesel tests, fuel hydrocarbons or aromatic volatiles were not detected in the treated soil. A semivolatile analysis of treated soil found approximately 30 ppm of unknown alkanes, benzoic acid, and other constituents. The results of metals analyses were similar to those of analysis of gasoline treated soils. Scrubber blowdown analyses revealed only a few compounds above detection limits, but none was at such a high level that would cause concern.

5.2 *Pilot-Scale Systems*

5.2.1 Petroleum Refinery Waste Sludge — Indirect-Heated Desorber

The Chemical Waste Management pilot-scale X*TRAX™ has been used to process a variety of petroleum wastes (Ayen and Swanstrom 1992). The process is an indirect-fired rotary desorber with inert gas passing through at a temperature of 315°C (600°F). The pilot-scale system is a 61 cm (24 in) in diameter by 6.4 m (21 ft) long desorber with a capacity of 0.05 kg/sec (0.20 ton/hr) for a feed with 30% moisture (Swanstrom and Palmer 1990). The inert purge gas passes through a series of condensers prior to recycle; a small portion of the gas is passed through a carbon filter and vented (≈5%). Waste codes K048, Dissolved Air Flotation Sludge, K049, Slop Oil Emulsion Solids, and Heat Exchanger Bundle Sludge, K050, were treated. The moisture content was 49% by weight and the oil and grease content was 23%. Soil exit temperatures were in the 290-360°C (554-680°F) range. Table 5.3 (on page 5.10) lists the results of the pilot-plant study. As shown in this table, most of the treated product met the best demonstrated available control technology (BDAT) for K048-50. The BDAT standards for metals, namely chromium and nickel, were not met since the materials exceeded TCLP Leachate concentrations, as shown in the table. Further treatment of the residual stream would be necessary. The authors state that the condensed oil recovered could feasibly be recycled.

Table 5.3

Chemical Waste Management Pilot-Scale Plant
Petroleum Refinery Waste

Total Constituent Analysis (ppm)

Constituent	Feed	Treated Prod.	BDAT for K048 - 50
Anthracene	9.2	0.37	28.0
Benzene	BDL (5)	BDL (0.5)	14.0
Benzo[a]pyrene	BDL (100)[1]	BDL (2)	12.0
Chrysene	BDL (100)[1]	1.3	15.0
Di-n-butylphalate	69.0	BDL (2)	3.6
Ethylbenzene	40.0	BDL (0.5)	14.0
Naphthalene	19.0	1.4	42.0
Phenanthrene	44.0	1.9	34.0
Phenol	BDL (100)[1]	0.46	3.6
Pyrene	57.0	1.3	36.0
Toluene	29.0	8.7	14.0
Xylene(s)	203.0	BDL (0.50)	22.0
Cyanide	0.5	---	1.8

Note: [1] BDAT standards are lower than method detection level (BDL)

TCLP Leachate Concentration (mg/L)
(2 samples shown)

Component	Feed	Treated Prod.	BDAT for K048 - 50
Chromium	0.28	2.7, 0.24	1.7
Nickel	0.25	0.55, 0.26	0.20

The authors also conducted laboratory-scale tests for the same materials; these tests yielded similar results. Additional contaminated soils have also been tested and results are discussed in Swanstrom 1991; Swanstrom and Palmer 1990, 1991; and Romzick and Swanstrom 1991.

5.2.2 PAH Contaminated Soils — Indirect-Fired Desorber

International Technology (IT) Corporation's Rotary Thermal Apparatus was used to treat soils contaminated with waste materials, including coal tar, which resulted from manufactured gas plant (MGP) operations (Alperin, Groen, and Helsel 1992; Fox, Alperin, and Huls 1991; Helsel, Alperin, and Groen 1989). Coal tar contains numerous PAHs. Three soils were tested with moisture contents from 4% (Soil A) to 9% (Soil C). In addition, material from a creosote and coal tar-based wood treater site was studied (Alperin, Groen, and Helsel 1992; Lauch et al. 1991). This soil contained 11% moisture.

The IT facility has a 16.5-cm (6.5 in.) internal diameter and a 2-m (6.7 ft) heated length. A nitrogen purge is used at a rate of 0.06 m^3/min (2 ft^3/min). The purge gas is cooled and passed through a high-efficiency particulate filter and an activated-carbon adsorber.

See table 5.4 (on page 5.12) for the results of several tests of the MGP soils and of a test creosote-contaminated soil. The MGP soil results show that an increase in temperature decreased the residual concentration; however, there did not appear to be a strong correlation with residence time. The heat up in the facility is quite rapid; therefore, the effect of residence time would be difficult to determine. The variability in final PAH concentration for different residence times is probably a result of inhomogeneity in the feed material. The creosote-contaminated soil test was conducted at 550°C (1,020°F) for 10 minutes. All concentrations were below detection limits.

For the creosote soil, the offgases were also analyzed. The PAHs and aromatic compounds were found as well as some phenolics and volatiles which were not in the parent soils. All PAHs that were in the parent soil were detected in the offgas. Phenanthrene and fluoranthene were used to determine a material balance, since these constituents were present in largest quantity (see table 5.4 on page 5.12). The material balance closure was 63% and 68% for phenanthrene and fluoranthene, respectively.

5.3 *Bench-Scale Systems*

Many of the case studies discussed above reported numerous tests on the bench scale. Studies of other wastes can also be found in the documents cited.

In many cases, the bench-scale studies were conducted prior to the pilot- and/or full-scale studies.

International Technology Corporation also conducted numerous tests with a bench-scale tray test apparatus. See the above sources for details.

Table 5.4

IT Rotary Thermal Apparatus
PAH Contaminated Soils

Part A

Creosote	Concentration mg/kg Feed	Concentration mg/kg Product
Naphthalene	143.0	<0.016
2-Methylnaphthalene	180.0	<0.190
Acenaphthalene	38.0	<0.007
Acenaphthene	342.0	<0.210
Dibenzofuran	237.0	<0.081
Fluorene	388.0	<0.020
Phenanthrene	1,028.0	<0.034
Anthracene	415.0	<0.073
Fluoranthene	668.0	<0.010
Pyrene	580.0	<0.052
Benzo[a]anthracene	17.0	<0.023
Chrysene	158.0	<0.120
Benzo[b]fluoranthene	158.0	<0.047
Benzo[a]pyrene	77.0	<0.110
Indeno[1,2,3-cd]pyrene	25.0	<0.035
Dibenzo[a,h]anthracene	<1.0	<0.016
Benzo[g,h,i]perylene	26.0	<0.320
Total Identified PAHs	4629	<n.d.

Part B

MGP Soil	Temperature °C	Time at Temperature (min)	Concentration mg/kg Feed	Concentration mg/kg Product
A	300	5.2	2,107	85.40
A	300	9.4	2,107	140.60
A	400	5.0	2,107	9.86
A	400	8.7	2,107	0.97
B	300	5.2	1,999	69.40
B	400	9.2	1,999	22.00
B	300	4.9	1,999	7.31
B	400	8.3	1,999	0.50
C	300	9.2	366	79.80
C	350	9.0	366	12.20
C	400	8.7	366	3.41

Part A Reprinted by permission of Edward Alperin from "Thermal Desorption of PAH-Contaminated Soil" by E.S. Alperin, A. Groen, and R.W. Helsel presented at 1992 Industrial Pollution Control Conference, Atlanta, February 9-12. Copyright 1992 by Edward Alperin.

Part B Courtesy Hazardous Waste Research and Information Center and Gas Research Institute

5.3.1 PAH Contaminated Soils

The University of Utah has completed bench-scale studies on the same MGP soils that were run in the IT pilot-scale system (Lighty, Eddings et al. 1990). The studies were conducted with IT Corporation in an attempt to scale the bench-scale results to the pilot-scale facility using computer modeling. The data from these systems showed that PAHs were reduced below 0.1 ppm at a temperature of 400°C (750°F), consistent with the pilot scale data. The source cited above also details exhaust gas measurements conducted with a gas chromatograph/mass spectrometry technique. Attempts were made to use the results of attempts to use the bench-scale data to predict the full-scale data by scaling with a computer model; these results are reported in Lighty, Silcox, and Pershing (1990). The results were inconclusive because of the difficulty in measuring of the hydrocarbons and the differences in reporting the concentrations between the bench-scale and pilot-scale data. The bench-scale data were given in terms of total hydrocarbons, whereas the pilot-scale data were constituent specific. It was found that the specific constituents accounted for only a fraction of the total hydrocarbons present.

6 LIMITATIONS

6.1 Waste Matrix

Several factors must be considered when deciding whether a waste can be treated by thermal desorption, including:

- Type of Contaminant and solid — treatability testing will help in determining the necessary temperature and time at temperature to achieve cleanup standards. Volatility of contaminant might be important;
- Contamination level — explosive limits must not be exceeded within systems operating with excess oxygen, and fugitive emissions must be controlled in an ex situ process;
- Moisture content — the energy required to remove moisture is a significant portion of the energy required to operate the system. Moisture also affects the desorption kinetics of the solid/contaminant matrix;
- Particle size distribution — important when considering material handling of the feed and the residual material;
- Environmental impacts of residuals — a material balance approach must be applied to the whole treatment system, not only the thermal desorber. The ultimate fate of all streams must be addressed and determined in the treatment decision; and
- Metals — the fate of the metals in the system must be determined. Metals must be captured before discharge from the stack and immobilized in residuals before replacement or disposal.

6.2 Process Needs

The system must be able to provide the necessary time-at-temperature to achieve the cleanup criteria. Fuel, electricity, and process water sources must be available. In addition, the process design must provide for complete, environmentally-sound disposal or elimination of all residuals.

6.3 Risk Considerations

Since the thermal desorption process is ex situ, fugitive emissions must be controlled. In addition, procedures must be implemented to preclude cross contamination. Cross contamination occurs when clean media is placed by a pile of contaminated media. As with any high temperature process involving potential contact with hot surfaces or the potential for explosion, careful consideration must be given worker safety. General industrial safety considerations apply; rotating equipment, conveyers, etc., require safety conscious operation. Exposure of personnel to hazardous materials must be minimized, and proper protection provided.

6.4 Site Considerations

Utility requirements and the space available for the remediation are basic considerations. Space requirements vary among the processes. If the facility is to be located in a residential area, these considerations are particularly important, especially as they affect public acceptance. Getting permits is an issue, especially for stationary facilities. For both mobile and stationary systems, permit acquisition is dependent on the physical location and specific characteristics of the site. Topography and meteorology are also important considerations.

6.5 Reliability of Performance

The systems addressed in this monograph have demonstrated reliability in the field tests reported; however, limited data are available on long-term operations. Most reliability problems occur not within the desorption system, but in material handling. Feeds and ash must be sized, screened, and conveyed; these processes often represent a considerable challenge, depending upon the material and type of equipment employed.

6.6 Process Residues

The process design must account for all residues of the thermal desorption process — gas, liquids, and solids. The cleanup and/or disposal of these streams must be planned. For example, if dioxins are present or are formed, and if heavy metals are present and/or enriched in the fly ash, their disposition must be provided for in the treatment program, since the disposal of these streams will affect the cost of treatment.

6.7 Quality of Treated Material

The quality of the treated material must be such that the residue can be disposed of or returned to the ground. If the requirements for cleanup are stringent for a particular contaminant, a treatability test might show that low-temperature thermal desorption will not meet the requirements and that higher temperatures are warranted. In addition, if the material is to be used as a final cover that must support plant growth, nutrients may have to be added.

6.8 Regulatory Requirements

Remediation of hazardous contaminated media is highly regulated (see also the discussion of regulatory requirements in Section 3.3). Resource Conservation Recovery Act (RCRA) Subpart X and, possibly, Subpart O may apply if the medium is classified as a hazardous waste under the Act or if the remediation is carried out under RCRA corrective action. If the cleanup is regulated under the Comprehensive Environmental Response, Compensation, and Liability Act (CERCLA), regulations under the RCRA may be applied as the applicable or relevant and appropriate regulations (ARAR). Applicable or relevant and appropriate regulations are determined on a case-by-case basis.

The process of selecting a remedy for a site under CERCLA is well delineated. After a remedial investigation/ feasibility study (RI/FS), a record of decision (ROD) is written by the Regional EPA Remedial Project Manager, specifying a selected remedy. The ROD becomes the basis for the consent decree or administrative (106) order or is included in an attached statement of work.

Although low-temperature thermal desorbers may not generally be as highly regulated as incineration systems, there is an exception. The US EPA's regulation on treatment of debris notes that a thermal desorber is regulated under the RCRA "either as an incinerator (if the device is direct-fired or if the off-gas is burned in an afterburner) under subpart O of part 264 or 265, or as a thermal treatment unit under subpart X, part 264 or subpart P, part 265."[1] Thus, regulations governing thermal desorption processes may vary depending upon the location of the site, contaminants, waste matrix, and system being employed. Therefore, it behooves practitioners to work closely and *early* with regulatory personnel to ascertain regulatory requirements.

1. Federal Register, August 18, 1992, page 37,194, footnote 24

7
TECHNOLOGY PROGNOSIS

Properly designed and operated thermal desorption units offer a viable means for the remediation of contaminated soils, sludges, and sediments. The technology can be used to treat a variety of wastes. A large volume of data has been generated concerning its use in treating petroleum-contaminated wastes.

Thermal desorption itself is but part of the total remediation process; pre- and postprocessing requirements are equally important when considering thermal desorption. As explained in Chapter 6.0, the technology has limitations that need to be addressed.

A number of vendors are available to provide the systems discussed in this monograph. A list of vendors and consultants who replied to a request for information in 1992 is set forth in Appendix D.

7.1 Development and Demonstration Needs

The following aspects of the technology have been identified as needing further development and demonstration:

- scaling of pilot or bench data to a full-scale system;
- fate of metals; and
- dioxin formation.

Figure 7.1 (on page 7.2) illustrates the importance of effectively scaling the system using computer modeling. The temperature history in a small, pilot-scale kiln might be completely different than that in a full-scale unit for the same fill fractions and wall temperatures. As shown in the figure, the bed reaches temperature within 500 seconds for 5% fill fraction in the pilot-scale unit. In the full-scale unit, the bed reaches temperature in 1,800 seconds for the

Figure 7.1

Fill-Fraction Predictions for Full-Scale (1.5 kg/sec feed rate) and Pilot-Scale (0.044 kg/sec feed rate) Rotary Kilns at Constant Revolution Rate of 0.2 rpm and Kiln Wall temperature of 800˚C

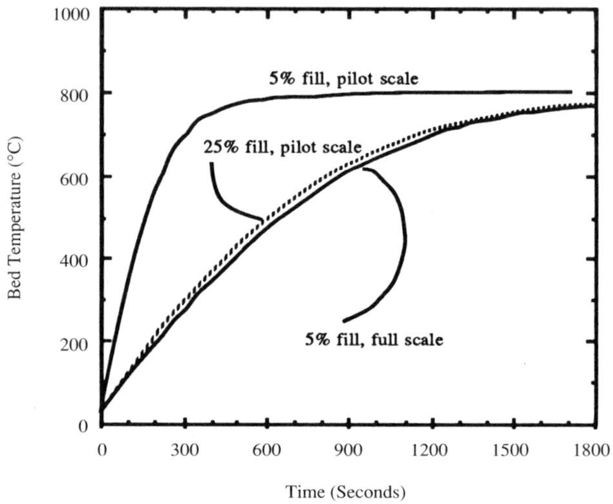

Reprinted by permission of the Combustion Institute From "Thermal Analysis of Rotary Kiln Incineration: Comparison of Theory and Experiment" by W.D. Owens, G.D. Silcox, J.S. Lighty, X.X. Deng, D.W. Pershing, V.A. Cundy, C.B. Ledger, and A.L. Jackway, Combustion and Flame, Volume 86, 1991. Copyright 1991 by the Combustion Institute.

same fill fraction. This difference is due to the heat transfer in the units. In fact, the pilot-scale unit would have to have a fill fraction of 25% to fully represent the same thermal profile as the full-scale unit. Lester and coworkers (1991) demonstrated some relation between full- and pilot-scale data even with a limited amount of full-scale data.

There is evidence that even at moderate temperatures, metals can be volatilized, especially volatile metals such as mercury. In the presence of chlorine, some chlorinated metal species could be formed that exhibit a more volatile nature than the parent oxide (Lighty, Eddings et al. 1990). In addition, metals can be volatilized and subsequently condensed on particles or condensed homogeneously to form small particulate which is enriched in metals (Barton, Clark, and Seeker 1990). This particulate could be difficult to capture in conventional air pollution control (APC) devices. If the metals are captured, disposal may be quite difficult, since they will probably not pass the Toxicity Characteristic Leachate Procedure (TCLP). Sorbent materials have been shown to capture some metals, and the metals react to form insoluble species (Uberoi and Shadman 1990, 1991; Scotto, Peterson, and Wendt 1992).

Two United States Environmental Protection Agency (US EPA) Superfund Innovative Technology Evaluation (SITE) demonstrations have produced evidence of dioxin/furan formation when solids contaminated with chlorinated aromatic organics were treated. These dioxins/furans are found in the intermediate liquid and gaseous treatment residuals, and are rarely detected in the treated soils/sediments. The mechanisms for the formation of dioxins and furans in these systems needs to be investigated further.

There is greater emphasis by US EPA regional offices and state regulatory personnel on the alternate uses of soils and sediments after thermal desorption treatment. Two factors affect the posttreatment usage: geophysical characteristics (e.g., strength), and toxicity. The geophysical characteristics are relatively easy to determine. Toxicity, however, is a very complex matter. Factors include not only carcinogenic, mutagenic, and lethal dose considerations, but also flora and fauna viability considerations. Procedures need to be developed for assuring that treated soil/sediment can meet minimum biological requirements for reuse.

APPENDIX A

Other Treatment Alternatives

Two other treatment systems were identified — fluidized beds and the Texarome Process — for which limited data is available relating to their application in thermal desorption as of August 1992. Therefore, they are addressed here rather than in the monograph proper.

The fluidized bed is a large refractor-lined vessel that can be divided into two sections: a plenum and a bed section. In the plenum section, air or hot gases pass through a distribution plate or tuyeres. The air rises up through the tuyeres and into the bed section. This air fluidizes a sand or granular-like material that constitutes the "bed." The bed serves as a heat transfer media. Extremely heavy fuel oils are lanced into the bed and are combusted directly as a source of auxiliary fuel. Natural gas and light oils are combusted in conventional burners, and the hot exhaust flows directly into the fluid bed to provide any required auxiliary heat. Particle diameters must be such that the medium can be fluidized. Also, the feed must not contain elements that form low-melting eutectics that would cause the bed to slag and defluidize. (See Roenzweig 1991, 9.)

The Texarome thermal desorption system is a continuous process using superheated steam as a conveying and stripping gas for treating contaminated materials. During the conveying process, the organic contaminants are separated from solid particles. An elaborate arrangement of piping within the conveying system allows for a countercurrent flow, as well as multistage dispersion and separation of the gas and solid phases. The last stage of the Texarome process is used as a quenching stage and a reactor loop to provide a chemical breakdown of the chemical residuals in the solids (US EPA 1992b). A SITE demonstration is being developed for a 6 kg/sec (25 ton/hr) unit. Presently a 3 kg/sec (12 ton/hr) pilot scale system is removing cedarwood oils from cedarwood chips. Steam temperatures vary from 150 to 480°C (350 to 900°F), depending on the contaminants. The system is to be transportable on three or four trailers or skids.

B

APPENDIX B

Engineering Bulletin: Thermal Desorption Treatment[1]

Purpose

Section 121(b) of the Comprehensive Environmental Response, Compensation, and Liability Act (CERCLA) mandates the Environmental Protection Agency (EPA) to select remedies that "utilize permanent solutions and alternative treatment technologies or resource recovery technologies to the maximum extend practicable" and to prefer remedial actions in which treatment "permanently and significantly reduces the volume, toxicity, or mobility of hazardous substances, pollutants and contaminants as a principal element." The Engineering Bulletins are a series of documents that summarize the latest information available on selected treatment and site remediation technologies and related issues. They provide summaries of and references for the latest information to help remedial project managers, on-scene coordinators, contractors, and other site cleanup managers understand the type of data and site characteristics needed to evaluate a technology for potential applicability to their Superfund or other hazardous waste site. Those documents that describe individual treatment technologies focus on remedial investigation scoping needs. This document is an update of the original bulletin published in May 1991 [1].

1 . US EPA. 1993. *Engineering Bulletin-Thermal Desorption on Treatment*. Vol. 2. EPA/540/0-00/000. OERR, Washington, D.C., and ORD, Cincinnati. In review.

Abstract

Thermal desorption is an ex situ means to physically separate volatile and some semivolatile contaminants from soil, sediments, sludges, and filter cakes by heating them at temperatures high enough to volatilize the organic contaminants. For wastes containing up to 10 percent organics or less, thermal desorption can be used in conjunction with offgas treatment for site remediation. It also may find applications in conjunction with other technologies or be appropriate to specific operable units at a site.

Thermal desorption is applicable to organic wastes and generally is not used for treating metals and other inorganics. The technology thermally heats contaminated media, generally between 200°F to 1000°F, thus driving off the water and volatile contaminants from the contaminated solid stream and transferring them to a gas stream. The contaminated gas stream is then treated by being burned in an afterburner, condensed to reduce the volume to be disposed, or captured by carbon adsorption beds.

The use of this well-established technology is a site-specific determination. Thermal desorption technologies are the selected remedies for one or more operable units at 31 Superfund sites [2]. Geophysical investigations and other engineering studies need to be performed to identify the appropriate measure or combination of measures to be implemented based on the site conditions and constituents of concern at the site. Site-specific treatability studies may be necessary to document the applicability and performance of a thermal desorption system. The EPA contact indicated at the end of this bulletin can assist in the definition of other contacts and sources of information necessary for such treatability studies.

This bulletin discusses various aspects of the thermal desorption technology including applicability, limitations of its use, residuals produced, performance data, site requirements, the status of the technology, and sources of further information.

Technology Applicability

Thermal desorption has been proven effective in treating organic-contaminated soils, sediments, sludges, and various filter cakes. Chemical contaminants for which bench-scale through full-scale treatment data exist include primarily volatile organic compounds (VOCs), semivolatile organic compounds (SVOCs), polychlorinated biphenyls (PCBs), pentachlorophenols (PCPs), pesticides, and herbicides [1][3][4][5][6][7]. The technology is not effective in separating inorganics from the contaminated medium.

Extremely volatile metals may be removed by higher temperature thermal desorption systems. However, the temperature of the medium produced by the process generally does not oxidize the metals present in the contaminated medium [8, p. 85]. The presence of chlorine in the waste can also significantly affect the volatilization of some metals, such as lead.

The technology is also applicable for the separation of organics from refinery wastes, coal tar wastes, wood-treating wastes, creosote-contaminated soils, hydrocarbon-contaminated soils, mixed (radioactive and hazardous) wastes, synthetic rubber processing wastes, paint wastes [4][9, p. 2] [10].

Performance data presented in this bulletin should not be considered directly applicable to other Superfund sites. A number of variables, such as concentration and distribution of contaminants, soil particle size, and moisture content, can all affect system performance. A thorough characterization of the site and a well-designed and conducted treatability study are highly recommended.

Table 1 lists the codes for the specific Resource Conservation and Recovery Act (RCRA) wastes that have been treated by this technology [4][9, p.7][10]. The indicated codes were derived from vendor data where the objective was to determine thermal desorption effectiveness for these specific industrial wastes.

The effectiveness of thermal desorption on general contaminant groups for various matrices is shown in Table 2. Examples of constituents within contaminant groups are provided in "Technology Screening Guide for Treatment of CERCLA Soils and Sludges" [8, p. 10]. This table is based on the current available information or professional judgement where no information was available. The proven effectiveness of the technology for a particular site or waste does not ensure that it will be effective at all sites or that the treatment efficiencies achieved will be acceptable at other sites. For the ratings used for this table, demonstrated effectiveness means that, at some scale, treatability was tested to show the technology was effective for that particular contaminant and medium. The ratings of potential effectiveness or no expected effectiveness are both based upon expert judgement. Where potential effectiveness is indicated, the technology is believed capable of successfully treating the contaminant group in a particular medium. When the technology is not applicable or will likely not work for a particular combination of contaminant group and medium, a no expected effectiveness rating is given.

Another source of general observations and average removal efficiencies for different treatability groups is contained in the Superfund Land Disposal Restrictions (LDR) Guide #6A, "Obtaining a Soil and Debris Treatability Variance for Remedial Actions," (OSWER Derective 9347.3-06FS, September 1990)[11] and Superfund LDR Guide #6B, "Obtaining a Soil and Debris Treatability Variance for Removal Actions," (OSWER Derective 9347.3-06BFS, September 1990)[12].

A further source of information is the U.S. EPA's Risk Reduction Engineering Laboratory Treatability Database (accessible via ATTIC).

Technology Limitations

Inorganics constituents and/or metals that are not particularly volatile will likely not be effectively removed by thermal desorption. If there is a need to remove a portion of them, a vendor process with a very high bed temperature is recommended; due to the fact that a higher bed temperature will generally result in a greater volatilization of contaminants. If chlorine or another chlorinated compound is present, some volatilization or inorganic constituents in the waste may also occur [13, p.8].

Table 1

RCRA Codes for Wastes Treated by Thermal Desorption

Wood Treating Wastes	K041
Dissolved Air Flotation	K048
Stop Oil Emulsion Solids	K049
Heat Exchanger Bundles Cleaning Sludge	K050
American Petroleum Institute (API) Separator Sludge	K051
Tank Bottoms (leaded)	K052

Table 2

Effectiveness of Thermal Desorption on General Contaminated Groups for Soil, Sludge, Sediments, and Filter Cakes

Contaminants Groups		Effectiveness: Soil	Sludge	Sedi-ments	Filter cakes
Organic	Halogenated volatiles	■	▼	▼	■
	Halogenated semivolatiles	■	▼	▼	■
	Nonhalogenated volatiles	■	▼	▼	■
	Nonhalogenated semivolatiles	■	▼	▼	■
	PCBs	▼	▼	▼	▼
	Pesticides	▼	▼	▼	▼
	Dioxins/Furans	▼	▼	▼	▼
	Organic cyanides	▼	▼	▼	▼
	Organic corrosives	❑	❑	❑	❑
Inorganic	Volatile metals	▼	▼	▼	▼
	Non volatile metals	❑	❑	❑	❑
	Asbestos	❑	❑	❑	❑
	Radioactive materials	❑	❑	❑	❑
	Inorganic corrosives	❑	❑	❑	❑
	Inorganic cyanides	❑	❑	❑	❑
Reactive	Oxidizers	❑	❑	❑	❑
	Reducers	❑	❑	❑	❑

■ Demonstrated Effectiveness: Successful treatability test at some scale completed

▼ Potential Effectiveness: Expert opinion that technology will work

❑ No Expected Effectiveness: Expert opinion that technology will not work

The contaminated medium must contain at least 20 percent solids to facilitate placement of the waste material into the desorption equipment [3, p.9]. Some systems specify a minimum of 30 percent solids [14 p.6].

As the medium is heated and passes through the kiln or desorber, energy is lost in heating moisture contained in the contaminated soil. A very high moisture content may result in low contaminant volatilization a need to recycle the soil through the desorber, or a need to dewater the material prior to treatment to reduce the energy required to volatilize the water. High moisture content, therefore, causes increased treatment costs.

Material handling of soils that are tightly aggregated or largely clay, or that contain rock fragments or particles greater than 1-1.5 inches can result in poor processing performance due to caking. The solids may have to be prepared by being crushed, screened, or shredded in order to meet the minimum treatment size. However, one advantage to soil preparation is that it the contaminated media is mixed and exhibits a more uniform moisture and BTU content.

If a high fraction of fine silt or clay exists in the matrix, fugitive dusts will be generated [8, p. 83] and a greater dust loading will be placed on the downstream air pollution control equipment [14, p.6].

The treated medium will typically contain less that 1 percent moisture. Dust can easily form in the transfer of the treated medium from the desorption unit, but can be mitigated by water sprays. Normally, clean water from air pollution control devices can be used for this purpose. Some type of enclosure may be required to control fugitive dust if water sprays are not effective.

Although volatile and semivolatile organics are the primary target of the thermal desorption technology, the total organic loading is limited by some systems to up to 10 percent or less [15, p. 11-30]. As in most systems that use a reactor or other equipment or process wastes, a medium exhibiting a very high pH (greater than 11) or very low pH (less than 5) may corrode the system components [8, p. 85].

There is evidence with some system configurations that polymers may foul and/or plug heat transfer surfaces [3, p.9]. Laboratory/field tests of thermal desorption systems have documented the deposition of insoluble brown tars (presumably phenolic tars) on internal system components [15, p. 76].

Caution should be taken regarding the disposition of the treatment material, since treatment processes may alter the physical properties of the material. For example, this material could be susceptible to such destabilizing forces as liquefaction, where pore pressures are able to weaken the material on sloped areas or places where materials must support a load (i.e., roads for vehicles, subsurfaces of structures, etc.). To achieve or increase the required stability of the treated material, it may have to be mixed with other stabilizing materials and/or compacted in multiple lifts. A thorough geotechnical evaluation of the treated product can provide the necessary design characteristics needed in order to achieve post-treatment soil stabilization [13, p.8].

There is also a concern that during the cleanup process dioxins and furans may form and be released from the exhaust stack and into the environment. Therefore, the system must be frequently monitored in order to ensure the unit is not releasing pollutants that may be harmful to the public and environment.

Technology Description

Thermal desorption is a process that uses either an indirect or direct heat exchange to heat organic contaminants to a temperature high enough to volatilize and separate them from a contaminated solid medium. Air, combustion gas, or an inert gas is used as the transfer medium for the vaporized components. Thermal desorption systems are physical separation processes, that transfer contaminants from one phase to another, and are not designed to provide high levels of organic destruction, although the higher temperatures of some systems will result in localized oxidation and/or pyrolysis. Thermal desorption is not incineration, since the destruction or organic contaminants is not the desired result. The bed temperatures achieved and residence times designed into thermal desorption systems will volatilize selected contaminants, but typically not oxidize or destroy them. System performance is typically measured by the comparison of untreated solid contaminant levels with those of the processed solids. The contaminated media is typically heated to 200°F to 1000°F, based on the thermal desorption system selected.

Figure 1 is a general schematic of the thermal desorption process.

Material handling (1) requires excavation of the contaminated solids or delivery of filter cake to the system. Typically, large objects greater than 1.5 inches are screened, crushed, or shredded and, if still too large, rejected. The medium is then delivered by gravity to the desorber inlet or conveyed by augers to a feed hopper [6, p.1].

Volatilization of contaminants can be effected by use of a rotary dryer, thermal screw, vapor extractor (fluidized bed), or distillation chamber [14].

As the waste is heated, the contaminants reach their respective boiling points, vaporize, and then become part of the gas stream. An inert gas, such as nitrogen, may be injected in a countercurrent sweep stream to prevent contaminant combustion and to aid in vaporizing and removing the contaminants [4][9,

Figure 1
Schematic Diagram of Thermal Desorption

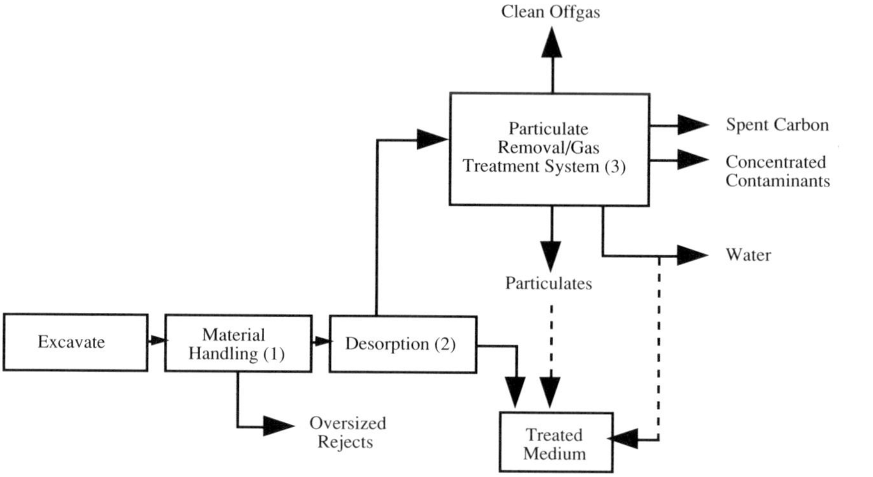

p. 1]. Other systems simply direct the hot gas stream from the desorption unit [3, p. 5][5].

The actual bed temperature and residence time are primarily factors affecting performance in the desorption stage. These factors are controlled in the desorption unit by using a series of increasing temperature zones [9, p. 1], multiple passes of the medium through the desorber where the operating temperature is sequentially increased, separate compartments where the heat transfer fluid temperature is higher, or sequential processing into higher temperature zones [16][17]. Heat transfer fluids used include hot combustion gases, hot oil, steam, and molten salts.

Offgas from desorption is typically processed (3) to remove particulates that remained in the gas stream after the desorption step. Volatiles in the offgas may be burned in an afterburner, collected on activated carbon, or recovered in condensation equipment. The selection of the gas treatment system will depend on the concentrations of the contaminants, cleanup standards, and the economics of the offgas treatment system(s) employed. Some methods commonly used to remove the particulates from the gas stream are a cyclone furnace, a quench water tower, and a baghouse. In a cyclone, particulates are removed by centrifugal force. In the quench tower, particulates are passed through a highly atomized water mist, causing them to discharge through the bottom of the tower. In a baghouse, particulates are caught by bags and discharged out of the system.

Process Residuals

Operation of the thermal desorption systems typically creates up to six process residual streams: treated medium oversized medium and debris rejects, condensed contaminants and water, particulate control system dust, clean offgas, and spent carbon (if used). Treated medium, debris, and oversized rejects may be suitable for return onsite.

Offgas from a thermal desorption unit will contain entrained dust (particulates) from the medium, vaporized contaminants, and water vapor. Particulates are removed by conventional equipment such as cyclone dust collectors, fabric filters, or wet scrubbers. Collected particulates may be recycled through the thermal desorption unit or blended with the treated medium, depending on the amount of carryover contamination present. Very small particles (<1 micron) can cause a visible plume from the stack [13, p.5].

The vaporized organic contaminants can be captured by condensing the offgas and then passing it through a carbon adsorption bed or other treatment system. Emissions may also be destroyed by use of an offgas combustion chamber or a catalytic oxidation unit [13, p.5].

When offgas is condensed, the resulting water stream may contain significant contamination depending on the boiling points and solubility of the contaminants and may require further treatment (i.e., carbon adsorption). If the condensed water is relatively clean, it may be used to suppress the dust from the treated medium. If carbon adsorption is used to remove contaminants from the offgas or condensed water, spent carbon will be generated, which is either returned to the supplier for reactivation/incineration or regenerated onsite [13, p.5].

When offgas is destroyed by a combustion process, compliance with incineration emission standards may be required. Obtaining the necessary permits and demonstrating compliance may be advantageous, however, since the incineration process would not leave residuals requiring further treatment. If incineration is used, the heat from the incineration process may be used in the desorption process unit [13, p.5].

Site Requirements

Thermal desorption systems are transported typically on specifically adapted flatbed semitrailers. Since most systems consist of three components (desorber, particulate control, and gas treatment), space requirements on site are typically less than 50 feet by 150 feet, exclusive of materials handling and decontamination areas.

Standard 440V, three-phase electrical service is needed. Water must be available at the site. The quantity of water needed is vendor and site specific.

Treatment of contaminated soils or other waste materials require that a site safety plan be developed to provide for personnel protection and special handling measures. Storage should be provided to hold the process product streams until they have been tested to determine their acceptability for disposal or release. Depending upon the site, a method to store waste that has been prepared for treatment may also be necessary. Storage capacity will depend on waste volume. Onsite analytical equipment capable of determining site-specific organic compounds for performance assessment make the operation more efficient and provide better information for process control.

Performance Data

Performance data in this bulletin are included as a general guideline to the performance of the thermal desorption technology and may not always be directly transferrable to other Superfund sites. Thorough site characterization and treatability studies are essential in determining the potential effectiveness of the technology at a particular site. Most of the data on thermal desorption come from studies conducted for EPA's Risk Reduction Engineering Laboratory under the Superfund Innovative Technology Evaluation (SITE) Program.

Seaview Thermal Systems (formerly T.D.I. Services, Inc.) conducted a pilot-scale test of their HT-5 thermal desorption system at the U.S. DOE, Oak Ridge, Tennessee, Y-12 plant. The test was run in order to evaluate the capability of the unit to remove and recover mercury from a soil matrix. Initial mercury concentrations in the soil were 1140 mg/kg. The mercury was removed to concentrations of 0.19mg/kg with a detection limit of 0.03 mg/kg. A full-scale cleanup (80 tons per day) using the HT-5 system, was conducted for Chevron U.S.A. at their El Segundo Refinery. The final primary contaminants and their initial and final concentrations are indicated in Table 3 [19].

Table 3

Full-Scale Cleanup Results of the H-T-5 System

Contaminant	Feed Soil Concentration (mg/kg)	Treated Soil Concentration (ug/kg)	Removal Efficiency [1] (%)
Tolune	30	< 620	< 97.93
Benzene	38	< 620	< 98.36
Ethlybenzene	93	< 620	< 99.79
Xylenes	290	< 620	< 99.78
Naphthalene	550	< 620	< 99.89
2-Methylnaphthalene	1400	< 330	< 99.98
Acenaphthlene	57	< 330	< 99.42
Phenanthrene	320	< 330	< 99.90
Anthracene	320	< 330	< 99.90
Pyrene	38	< 330	< 99.13
Benzo(a) Anthracene	36	< 330	< 99.08
Chrysene	45	< 330	< 99.27
Styrene	13	< 620	< 99.23

1. SIC Re: <

In September 1992, a pilot-scale (35 ton per hour capacity) EPA Superfund Innovative Technology Evaluation (SITE) demonstration was performed at a confidential Arizona pesticide site using Canonie Environmental's Low Temperature Thermal Aeration (LTTA®) system. Approximately 1,180 tons of pesticide-contaminated soil was treated during the demonstration over three 10-hour replicate runs. Pesticide-contaminated soils ranged in concentration from 7,080 ug/kg to 1,540,000 ug/kg. The LTTA® system obtained removal efficiencies ranging from 82.4 to greater than 99.9 percent. All pesticides with the exception of dichlorodipheyltrichloroethane (DDE), were removed to near or below method detection limits in the soil. Trace amounts of polycholorinated dibenzo(p)dioxins (PCDD) and polychlorinated dibenzofurans (PCDF) were formed in the LTTA® system but at concentrations too low to quantify. No Class I or Class II dioxin precursors were found above detection limits. Total chloride concentrations in the treated soil were approximately three times the total chloride concentrations in the feed soil. This increase was likely due to the dechlorination of the pesticides in the feed soil. Table 4 Presents a summary of four case studies involving full-scale applications of the LTTA® process [17].

Table 4

Full-Scale Cleanup Results of the LTTA® System

Site	*Volume/Mass Treated*	*Primary Contaminant(s)*	*Feed Soil Concentration (mg/kg)*	*Treated Soil Concentration (mg/kg)*
South Kearney	16,000 tons	Total VOCs	308.2	0.51
		SVOCs	0.7-15	ND-1.0
McKin	11,500 cubic yards	VOCs	2.7-3,310	<0.05[a]
		SVOCs	0.44-1.2	<0.33-0.51
Ottati and Goss	4,500 cubic yards	1,1,1 TCA	12-470	<0.025
		TCE	6.5-460	<0.025
		Tetrachloroethene	4.9-1200	<0.025
		Toluene	>87-3,000	<0.025-0.011
		Ethylbenzene	>50-440	<0.025
		Total Xylenes	>170->1100	<0.025-0.014
Cannon Bridgewater	11,300 tons	VOCs	5.30[b]	<0.025
Former Spencer Kellogg Facility	6,500 tons	Total VOCs	5.42	0.45
		SVOCs	0.15-4.7	0.042-<0.39

a Average concentration
b Maximum concentration

A pilot-scale (2.1 tons per hour capacity) SITE demonstration was performed at the Anderson Development Company (ADC) Superfund site in Adrian, Michigan using Roy F. Weston's Low Temperature Thermal Treatment (LT^3®) system. Approximately 80 tons of contaminated sludge was treated during the demonstration over a period of six 6-hour replicate tests. The lagoon sludge was primarily contaminated with VOCs and SVOCs, including 4,4'-

methylenebis (2-chloroanilinine) (MBOCA). Initial VOC concentrations ranged from 35 to 25,000 ug/kg. In the treated sludge VOC concentrations were below method detection limits (less than 60 ug/kg) for most compounds. MBOCA concentrations in the untreated sludge ranged from 43.6 to 960 mg/kg. The treated sludge ranged in concentration from 3 to 9.6 mg/kg. The LT^3® system also decreased the concentrations of all SVOCs present in the sludge, with two exceptions: chrysene and phenol. The increase of chrysene concentration likely was caused by a minor leak of heat transfer fluid. Chemical transformations during heating likely caused the phenol concentrations to increase. PCDDs and PCDFs were formed in the system, but were removed from the exhaust gas by the unit's vapor-phase carbon column with removal efficiencies, varying with congener, from 20 to 100 percent. Particulate concentrations ranged from less than 8.5×10^{-4} to 6.6×10^{-3} grains per dry standard cubic meter (gr/dscm). Chloride concentrations were below the average method detection limit of 2.8×10^{-2} milligrams per dry standard cubic meter (mg/dscm). Chloride emissions were less than 6.0×10^{-5} lb/hr. Table 5 presents a summary of three case studies involving pilot- and full-scale applications of the LT^3® system [21].

Table 5

Full-Scale Cleanup Results of the LT^3® System

Site	*Volume/Mass Treated*	*Primary Contaminant(s)*	*Feed Soil Concentration*	*Treated Soil Concentration*
Confidential	1,000 cubic feet	Benzene	1,000 ppb	5.2 ppb
		Toluene	24,000 ppb	5.2 ppb
		Xylene	110,000 ppb	<1.0 ppb
		Ethylbenzene	20,000 ppb	4.8 ppb
		Napthalene	4,900 ppb	<330
		PAHs	890-<6,000 ppb	<330 - 590 ppb
Tinker AFB, OK	3,000 cubic yards	TCE Chlorinated Organics JP4 Aviation Fuel	6,100 ppma	All contaminants were reduced to concentrations below the goal cleanup levels and in most cases, to less than detection limits
Letterkenny Army Depot	7.5 tons	Benzene	586,106 ppb	730 ppb
		Trichloroethene	2,678,536 ppb	1,800 ppb
		Tetrachloroethene	1,422,031 ppb	1,400 ppb
		Xylene	27,197,367 ppb	550 ppb
		Other VOCs	39,127 ppb	BDL

a Maximum concentration
BDL Below Detection Limits

In May 1991 a pilot-scale SITE demonstration was performed at the Wide Beach Development site in Brand, New York using Soil Tech's Anaerobic Thermal Processor (ATP) system. Approximately 104 tons of contaminated soil were treated during three replicate test runs. The soil and sediment at the site was primarily contaminated with PCBs, along with VOC's and SVOCs. The average total PCB concentration was reduced from 28.2 mg/kg, in the contaminated soil and sediment, to 0.043 mg/kg in the treated soil (a 99.8 percent removal efficiency). The test indicated that an average of 23.1 ug/dscm of PCBs were discharged from the unit's stack to the atmosphere. The high PCB concentrations in the emissions may have been caused by low removal efficiencies in the unit's vapor phase carbon system, high particulate loadings (0.467 g/dscm) in the stack or a combination of the two. Low levels of dioxins and furans were present in the feed soil, but none were detected in the treated soils, baghouse fines, or the cyclone's flue gas. The toxicity equivalents (TEQ) of the stack gas ranged from 0.0106 to 0.0953 mg/dscm. The total VOC concentration was reduced from 0.085 mg/kg in the contaminated soil to 0.0008 mg/kg in the treated soil. The total SVOC concentration was reduced from 61.8 mg/kg in the contaminated soil to 0.24 mg/kg in the treated soil [22].

In June 1991 a SITE demonstration test was performed at the Waukegan Harbor Superfund site in Waukegan Harbor, IL. The site was primarily contaminated with PCBs, along with VOCs, SVOCs, and metals. Approximately 253 tons of contaminated soil were treated over a period of four runs using Soil Tech's ATP thermal desorption system. The average PCV concentration in the feed soil was 9,173 mg/kg; the average final concentration was 2 mg/kg, which is a 99.98 percent removal efficiency. PCB's were discharged out of the stack at 0.834 ug/dscm (a 99.999987 percent removal efficiency). Tetrachlorinated dibenzofurans were the only dioxins and furans detected in the stack gas at an average concentration of 0.0787 ng/dscm. Low concentrations of SVOCs (total of 16.99 mg/kg) in the feed soil were detected. In the treated soils SVOC concentrations totaled only 0.031 mg/kg with only two contaminats detected below the concentration limit. In the contaminated soil, VOC concentrations totaled 17 mg/kg, while in the treated soil the total was only 0.03 mg/kg. Metal concentrations were approximately the same in both the contaminated and treated soil. This was due to the fact that the unit does not operate at temperatures high enough to significantly remove metals. The pH of the soil rose from 8.59 in the contaminated soil to 11.35 in the treated soil. This was likely to the addition of sodium bicarbonate in order to reduce PCB emissions [22].

In May 1992 a pilot-scale (4.9 tons per hour) SITE demonstration was performed at the RE-Solve Superfund site in North Dartsmouth, Massachusetts using Chemical Waste Management X*TRAX™ system. Approximately 215 tons of contaminated soil were treated over a period of three duplicate six-hour tests. The soil is primarily contaminated with PCBs, along with some oil and grease, and metals. Initial PCB concentrations ranged from 181 to 515 mg/kg. PCB concentrations in the treated soil were less than 1.0 mg/kg with the average concentrations being 0.25 mg/kg (a 99.9 percent removal efficiency). PCDDs and PCDFs were not formed during the demonstration. Concentrations of oil and grease, total recoverable petroleum hydrocarbons, and tetrachloroethane were reduced to below detectable levels. Metal concentrations were not reduced during the test. This was expected since the unit does not operate at temperatures high enough to significantly remove metals [23].

RCRA LDRs that require treatment of wastes to best demonstrated available technology (BDAT) levels prior to land disposal may sometimes be determined to be applicable or relevant and appropriate requirements for CERCLA response actions. Thermal desorption can produce a treated waste that meets treatment levels set by BDAT but may not reach these treatment levels in all cases. The ability to meet required treatment levels is dependent upon the specific waste constituents and the waste matrix. In cases where thermal desorption does not meet these levels, it still may, in certain situations, be selected for use at the site if a treatability variance establishing alternative treatment levels is obtained. Treatability variances are justified for handling complex soil and debris matrices. The following guides describe when and how to seek a treatability variance for soil and debris: Superfund LDR Guide #6 "Obtaining a Soil and Debris Treatability Variance for Remedial Actions" (OSWER Directive 9347.3-06FS) [11], and Superfund LDR Guide #6B, "Obtaining a Soil and Debris Treatability Variance for Removal Actions" (OSWER Directive 9347.3-06BFS)[12].

Technology Status

Several firms have experienced in implementing this technology. Therefore, there should not be significant problems of availablity. The engineering and configuration of the systems are similarly refined, such that once a system is designed full-scale, little or no prototyping or redesign is required.

A SITE demonstration is scheduled to take place in June 1993 at the Niagara Mohawk Power Corporation site in Utica, New York. The facility is a former gas manufacturing plant and contains 425,000 cubic yards of manufactured gas plant soil. The soil is primarily contaminated with polyaromatic hydrocarbons (PAHs), benzene, toluene, ethylbenzene, xylene (BTEXs), lead, arsenic, cyanide. EPA Technology Evaluation and Application Analysis Reports will be developed in order to evaluate the performance of and the cost to implement the system.

Thermal desorption technologies are the selected remedies for one or more operable units at 31 Superfund sites. Table 6 presents the status of selected Superfund sites employing the thermal desorption technology [2].

Table 6

Superfund Sites Specifying Thermal Desorption as the Remedial Action

Site	*Location (Region)*	*Primary Contaminant(s)*	*Status*
Cannon Engineering (Bridgewater Site)	Bridgewater, MA (1)	VOCs (Benzene, TCE, Toluene, Vinyl Chloride)	Project completed 10/90
McKin	McKin, ME (1)	VOCs, (TCE, BTX)	Project completed 2/87
Ottati & Goss	New Hampshire (1)	VOCs, (TCE, PCE, 1,2-DCE, Benzene)	Project completed 9/89
Wide Beach Development	Brandt, NY (2)	PCBs	Pilot study completed 5/91
Metaltec/Aerosystems	Franklin Borough, NJ (2)	VOCs (TCE)	Design completed
Caldwell Trucking	Fairfield, NJ (2)	VOCs (TCE, PCE, TCA)	Design completed
Outboard Marine/Waukegan Harbor	Waukegan Harbor, IL (5)	PCBs	Pilot study completed 6/92
Reich Farms	Dover Township, NJ (2)	VOCs (TCE, PCE, TCA), SVOCs	Pre-design
Re-Solve	North Dartmouth, MA (1)	PCBs	Pilot study completed 5/92
Waldick Aerospace Devices	New Jersey, (2)	VOCs (TCE, PCE), Metals (Cadium, Chromium)	Design completed
Waumchem	Burton, SC (4)	VOCs (BTX)	In design
Fulton Terminals	Fulton, NJ (2)	VOCs, (Xylene, TCE, Benzene, DCE)	In design
Anderson Development Company	Adrian, MI (5)	VOCs, SVOCs	Pilot study completed 12/91

NOTE: The two Stauffer Chemical sites in Table 10 of the first version are not included in this table due to the fact that EPA's ROD Annual Report FY 1990 indicates that thermal desorption will no longer be implemented.

Several vendors have experience in the construction of this technology and have documented processing costs per ton of feed processed. The overall range varies from approximately $50 to $400 (1993 dollars) per ton processed [22]. Caution is recommended in using costs out of context because the base year of the estimates vary. Costs also are highly variable due to the quantity of waste to be processed, term of the remediation contract, moisture, content, organic constituency of the contaminated medium, and cleanup standards to be achieved. Similarly, cost estimates should include such items as preparation of Work Plans, permitting, excavation, processing itself, QA/QC verification of treatment performance, and reporting of data.

EPA Contacts

Technology-specific questions regarding thermal desorption may be directed to:

Paul dePercin
U.S. Environmental Protection Agency
Risk Reduction Engineering Laboratory
26 W. Martin Luther King Drive
Cincinnati, Ohio 45268
(513) 569-7797

James Yezzi
U.S. Environmental Protection Agency
Risk Reduction Engineering Laboratory
Releases Control Branch
2890 Woodbridge Avenue
Building 10 (MS-104)
Edison, NJ 08837
(908) 321-6703

Acknowledgments

This updated bulletin was prepared for the U.S. Environmental Protection Agency, Office of Research and Development (ORD), Risk Reduction Engineering Laboratory (RREL), Cincinnati, Ohio, by Science Applications International Corporation (SAIC) under Contract No. 68-CO-0048. Mr. Eugene Harris served as the EPA Technical Project Monitor. Mr. Jim Rawe (SAIC) was the Work Assignment Manager. He and Mr. Eric Saylor (SAIC) co-authored the revised bulletin. The authors are especially grateful to Mr. Paul dePercin of EPA RREL, who contributed significantly by serving as a technical consultant during the development of this document.

The following other Agency and contractor personnel have contributed their time and comments by participating in the expert review meetings and/or peer reviewing the document:

Dr. James Cudahy	Focus Environmental, Inc.
Dr. Steven Lanier	Energy and Environmental Research Corp.

References

1. Thermal Desorption Treatment. Engineering Bulletin. U.S. Environmental Protection Agency, EPA/540/2-91-008. May 1991.
2. Innovative Treatment Technologies. Semi-Annual Status Report (Fourth Edition), U.S. Environmental Protection Agency. EPA/542/R-92/011.
3. Abrishamian, Ramin. Thermal Treatment of Refinery Sludges and Contaminated Solids. Presented at American Petroleum Institute, Orlando, Florida, 1990.
4. Swanstrom, C., C. Palmer. X*TRAX® Transportable Thermal Separator for Organic Contaminated Solids. Presented at Second Forum on Innovative Hazardous Waste Treatment Technologies: Domestic and International, Philadelphia, Pennsylvania, 1990.
5. Canonie Environmental Services Corp., Low Temperature Thermal Aeration (LTTA℠) Marketing Brochures, circa 1990.
6. VISITT Database, 1993.

7. Nielson, R., and M. Cosmos. Low Temperature Thermal Treatment (LT3) of Volatile Organic Compounds from Soil: A Technology Demonstrated. Presented at the American Institute of Chemical Engineers Meeting, Denver, Colorado, 1988.

8. Technology Screening Guide for Treatment of CERCLA Soils and Sludges. EPA/540/2-88/004, U.S. Environmental Protection Agency, 1988.

9. T.D.I. Services, Marketing Brochures, circa 1990.

10. Cudahy, J., W. Troxler. 1990. Thermal Remediation Industry Update - II. Presented at Air and Waste Management Association Symposium on Treatment of Contaminated Soils, Cincinnati, Ohio, 1990.

11. Superfund LDR Guide #6A: (2nd Edition) Obtaining a Soil and Debris Treatability Variance for Remedial Actions. Superfund Publications 9347.3-06FS, U.S. Environmental Protection Agency, 1990.

12. Superfund LDR Guide #6B: Obtaining a Soil and Debris Treatability Variance for Remedial Actions. Superfund Publications 9347.3-06FS. U.S. Environmental Protection Agency, 1990.

13. Guide for Conduction Treatability Studies under CERCLA Thermal Desorption Remedy Selection, Interim Guidance, U.S. Environmental Protection Agency. EPA/540/R-92-074A.

14. Recycling Sciences International, Inc., DAVES Marketing Brochures, circa 1990.

15. The Superfund Innovative Technology Evaluation Program - Progress and Accomplishments Fiscal Year 1989. A Third Report to Congress, EPA/540/5-90/001, U.S. Environmental Protection Agency, 1990.

16. Superfund Treatability Clearinghouse Abstracts. EPA/540/2-89/001, U.S. Environmental Protection Agency, 1989.

17. Soil Tech, Inc., AOSTRA - Tuciuk Processor Marketing Brochure, circa 1990.

18. Ritcey, R., and F. Schwarz, Anaerobic Pyrolysis of Waste Solids and Sludges - The AOSTRA Taciuk Process System. Presented at Environmental Hazardous Conference and Exposition, Seattle, Washington, 1990.

19. Seaview Thermal Systems, Marketing Brochures, circa 1993.

20. Low Temperature Thermal Treatment Aeration (LTTA) Technology.

Canonie Environmental Services Corporation. Applications Analysis Report, U.S. Environmental Protection Agency (Draft).

21. Roy F. Weston, Inc. Low Temperature Thermal Treatment (LT^3®) System. Applications Analysis Report. Anderson Development Company Site, U.S. Environmental Protection Agency (Draft).

22. Soil Tech ATP Systems, Inc. Anaerobic Thermal Processor. Applications Analysis Report. Wide Beach Development Site and Outboard Marine Corporation Site. U.S. Environmental Protection Agency (Preliminary Draft).

23. X*TRAX Model 200 Thermal Desorption System. Chemical Waste Management, Inc. Demonstration Bulletin. U.S. Environmental Protection Agency (Draft).

C

APPENDIX C

THERMAL DESORPTION OF PCB CONTAMINATED WASTE AT THE WAUKEGAN HARBOR SUPERFUND SITE

A CASE STUDY (Excerpt)[1]

Abstract

In June 1992, SoilTech ATP Systems, Inc. (SoilTech) completed the soil treatment phase of the Waukegan Harbor Superfund Project in Waukegan, Illinois, after approximately five months of operation. SoilTech successfully treated 12,700 tons of polychlorinated biphenyl (PCB) contaminated sediments using a transportable SoilTech Anaerobic Thermal Processor (ATP) System nominally rated at 10 tons-per-hour throughput capacity. The SoilTech ATP Technology anaerobically desorbs contaminants such as PCBs from solids and sludges at temperatures over 1,000 degrees Fahrenheit (°F). Principal products of the process are clean treated solids and an oil condensate containing the hydrocarbon contaminants.

At the Waukegan Harbor Superfund Site, PCB concentrations in the sediments excavated and dredged from a ditch, lagoon and harbor slip averaged

1. Excerpted from "Thermal Desorption of PCB Contaminated Waste at The Waukegan Harbor Superfund Site - A Case Study" by J.H. Hutton and R. Shanks presented at Innovative Treatment Technologies - Uses and Applications for Site Remediation, Thermal I - Thermally Enhanced Volatilization, Satellite Seminar, February 18, 1993. Reprinted by permission of the Air & Waste Management Association.

10,400 parts per million (ppm) (1.04 percent) and were as high as 23,000 ppm (2.3 percent). Treated soil contained less than 2 ppm PCBs and was backfilled in an on-site containment cell. The removal efficiency of PCBs from the soil averaged 99.98 percent, relative to the project performance specification of 97 percent. Approximately 30,000 gallons of PCB oil, desorbed from the feed material, were returned to the potentially responsible party (PRP) trust for subsequent off-site disposal. After modifications to the emissions control equipment, compliance with the 99.9999 percent destruction and removal efficiency (DRE) for PCBs in stack emissions required by the United States Environmental Protection Agency was achieved. In fact, SoilTech demonstrated compliance with the DRE requirement in eleven consecutive stack sampling events. Feed rate averaged 8 tons-per-hour at a mechanical availability of 85 percent. SoilTech revenues for the project were $700,000 in fixed costs and $185 per ton of soil processed.

1.0 INTRODUCTION

This technical paper outlines SoilTech's role in the Waukegan Harbor Superfund Project. SoilTech was responsible for the soils processing phase of the project using their unique rotary kiln known as the SoilTech Anaerobic Thermal Processor (ATP) Technology to remediate the PCB contaminated soils and sediments. The Waukegan Harbor Project is the second commercial application of the SoilTech ATP Technology. 42,000 tons of PCB contaminated soils were successfully treated to nondetect levels at the Wide Beach Superfund site in western New York in 1990 and 1991. The ATP distills organics contaminants out of a solid matrix in an oxygen free environment. Oxidative degradation of contaminants such as PCBs into more harmful reaction products is therefore prevented. Contaminants are collected in an oily condensate, which can then be economically disposed of.

4.0 WAUKEGAN HARBOR SUPERFUND PROJECT DESCRIPTION

The Waukegan Harbor Superfund site was listed on the EPA's Superfund Priority List. Contamination resulted from leakage of PCBs - used as an aluminum casting lubricant and machine tool lubricant - through floor drains from a major manufacturing facility into an adjacent stream and Waukegan Harbor. PCB concentrations in excess of 20,000 ppm were found in the harbor sediments and stream bed. Remediation of the site was complicated by the fact that the harbor was being used for both commercial and recreational boating activities and was adjacent to a public beach. The client was insistent that operations not impact the day-to-day activities of the local community. It was, therefore, necessary for Canonie, the prime contractor, to build a new boat slip prior to isolating the old one. Sediments with PCB concentrations greater than 500 ppm were hydraulically dredged from the old slip and pumped to a containment cell where dewatering was achieved before they were blended with soils from the contaminated stream bed.

The 1984 Record of Decision called for stabilization of these soils and sediments. SoilTech and the PRPs were, however, able to convince the EPA that the use of SoilTech's ATP Technology offered a more environmentally acceptable and cost-effective solution. The EPA accepted that the SoilTech ATP System was an innovative treatment technology that provided a significant and permanent reduction in the toxicity and volume of PCB wastes at the Waukegan Harbor Site. SoilTech's approach fulfilled the requirements of the Superfund Amendments and Re-authorization Act (SARA). Under the 1984 ROD, there would have been a potential future liability posed by stabilized stored wastes.

SoilTech was contracted by Canonie to process the soils and sediments for \$700,000 in fixed costs and \$185 per ton of material processed. SoilTech was not responsible for providing utilities (water, gas, and electricity). Site preparation and excavation of contaminated material were carried out by Canonie. The PRP's took full responsibility for disposing of the PCB condensate produced. Had SoilTech been responsible for any or all of these additional activities, the unit price for soils processing would have been correspondingly higher.

The project clean-up criteria called for a treated soil residual PCB level of 500 ppm, or 97 percent removal efficiency from the soils, whichever was more stringent. The average target clean-up level turned out to be approximately 310 ppm. Total soils requiring treatment amounted to about 12,700 tons. Treated soils were to be placed in one of two containment cells along with untreated soils containing less than 500 ppm PCBs before being closed with a TSCA approved impermeable cap. PCB contaminated oils extracted from the soils were returned to the PRP for subsequent disposal.

Since no applicable air emissions standard exists for thermal desorption, the EPA required that SoilTech meet the 99.9999 percent (6 nines) destruction and removal efficiency (DRE). SoilTech was also required to meet the municipal waste incineration standard of 30 ng/m^3 total dioxins and furans applied to incineration of PCB contaminated wastes. Discharged water was to be less than 1 part-per-billion (ppb) PCBs.

4.1 Sequence of Events

SoilTech arrived at the Waukegan Harbor Superfund site in the last week of November 1991. Modification and repair of key process components continued at a shop in Indiana while the site was prepared, and support equipment assembled and erected. Shakedown and troubleshooting of the SoilTech ATP System was conducted in mid-January. SoilTech began treating contaminated soils and sediments on January 22, 1992. This date also marked the start of the 30-day proof-of-process period.

Between startup and March 6, 1992 the SoilTech ATP System averaged a feed rate of 10 tph outperforming initial projections of 9 tph. Net plant availability was over 85 percent. PCB concentrations in the feed averaged 14,000 ppm and treated soil concentrations less than 2 ppm PCBs were attained. The average removal efficiency from the treated soil was 99.96 percent.

During this time period, SoilTech performed seven stack tests. Although the emissions criteria for dioxins and furans were met, SoilTech was unable to meet the 6 nines DRE for PCBs and consequently stopped operations at the direction of the EPA.

During the next two months, SoilTech operated the ATP System intermittently to test various modifications to the ATP and to the emission control system. Specific modifications included:

- Increased volume of carbon in flue gas carbon bed;
- Removal of the wet scrubber from the flue gas handling system;
- Addition of carbon beds to internal gas recycle streams;

- Testing of continuous addition of powdered activated carbon upstream of the baghouses to facilitate adsorption of PCBs;
- Reduction of flue gas carbon bed temperature; and
- Testing of continuous addition of sodium bicarbonate in the combustion zone of the ATP in an attempt to induce catalytic destruction of residual PCBs.

In the latter phase of this period of intermittent operation and testing, SoilTech discovered a gap in the flue gas carbon bed seal which was allowing 70 percent of the flue gas stream to bypass the carbon bed. SoilTech corrected the poor seal before stack testing on May 12, 13, and 14, 1992.

Each of four stack tests conducted on those three consecutive days demonstrated performance superior to the 6 nines DRE required. A summary of the stack testing conducted throughout the project is given in Table 1.

It appeared that the final three tests, using powdered activated carbon in the baghouse and then soda ash in the combustion zone, produced slightly better

Table 1

Stack Test Summary
Waukegan Harbor

Date	Feed PCBs (#/hr)	Stack PCBs (#/hr)	PCB DRE[1] (%)
01-28-92	192.50	0.0144000	99.9925195
02-04-92	215.50	0.0932000	99.9567517
02-11-92	213.40	0.0690000	99.9676664
02-18-92	155.70	0.0087600	99.9943738
03-04-92	105.60	0.0039700	99.9962405
03-05-92	100.80	0.0012700	99.9987401
03-05-92	100.80	0.0007720	99.9992341
03-18-92	174.90	0.0009890	99.9994345
04-09-92	148.60	0.0034500	99.9976783
04-10-92	298.50	0.0009430	99.9996841
05-12-92	230.40	0.0002000	99.9999132
05-13-92	264.60	0.0000735	99.9999722
05-13-92	183.00	0.0000464	99.9999746
05-14-92	205.70	0.0000448	99.9999782
06-02-92	167.40	0.0000942	99.9999437
06-02-92	213.60	0.0001850	99.9999134
06-09-92	159.15	0.0000432	99.9999729
06-16-92	140.80	0.0000050	99.9999964

Note:
1. PCB Destruction Removal Efficiency (DRE) criterion is 99.9999 percent.

results than the test conducted without these additives. The results are, however, statistically inconclusive.

With EPAs approval SoilTech was then able to resume processing the contaminated soils and sediments having met all of the project performance criteria.

To insure that this performance was maintained, SoilTech instituted several new procedures and operating conditions.

- Soda ash fed to combustion zone at 200 lbs/hour;
- Stack temperature maintained at 170°F with automatic waste feed shut off at 190°F;
- Sample the stack gas carbon bed on a daily basis. Shut down plant and change the carbon bed when the PCB concentration exceeds 10 ppm;
- Weekly stack testing to be performed by Clean Air Engineering. Failure to meet 6-nines DRE would necessitate immediate shutdown until the cause had been identified and remedied; and
- Monitor pressure drop across stack carbon bed. Inspect bed if it drops below 5-inches water column.

4.2 EPA Site Demonstration

On June 16, 1992 the EPA began a Superfund Innovative Technology Evaluation (SITE) demonstration of the SoilTech ATP System. This involved three days of rigorous sampling and testing. The test runs were conducted at typical operating conditions. The fourth and final test was, however, run without the addition of soda ash to the combustion zone. The other three tests were run while 200 lbs/hr or soda ash was added to the combustion zone. Each test run consisted of 8.5 hours of solids and liquids sampling and 8 hours of stack sampling. During the site demonstration, 224 tons of PCB contaminated soils and sediments were processed. In addition to sampling, critical operating parameters were collected during each test.

Based on the preliminary results at the site demonstration, the EPA concluded the following:

- PCB concentrations were reduced from an average of 9,761 parts per million (ppm) in the untreated soil and sediment to an average concentration of 2 ppm in the treated soil and sediment.

- Approximately 0.12 milligrams (mg) of PCBs were discharged from the ATP System's stack per kilogram of PCBs fed to the ATP.
- The majority of PCBs removed from the untreated soil and sediment were accumulated in the waste oil discharge from the vapor cooling system.
- No dioxins, other than a low concentration [0.1 nanograms (ng) per dry standard cubic meter (dscm)] of octachlorinated dibenzo-p-dioxin in one stack gas sample, were detected in the stack gas from the ATP System. Tetrachlorinated dibenzofurans were found in both the untreated soil and sediment (88 ng/g) and treated soil and sediment (5ng/g) and the stack gas (0.07 ng/dscm).
- Leachable VOCs, SVOCs, and metals in the treated soil and sediment were below Resource Conservation and Recovery Act (RCRA) toxicity characteristic standards.
- No operational problems affecting the ATPs ability to treat the contaminated soil and sediment were observed.

Selected results provided by the EPA are presented in Table 2 (on page C.8). As expected, the results confirm that SoilTech met all of the project performance criteria. Removal efficiency for PCBs in the treated soil averaged 99.98 percent. DRE for the stack emissions exceeded the 6 nines criteria in each test and bettered 7 nines in two of the tests.

The results also demonstrated that dioxins and furans are not produced in the ATP. Trace amounts detected in the waste oil, the treated soil and stack gas can be traced back to the untreated soils and sediments.

The highest DRE for PCBs in the stack was achieved for the test in which no soda ash was added to the combustion zone of the ATP. This indicates that the addition of soda ash provides no reduction in the flue gas emissions produced by the SoilTech ATP System. Further bench testing might, however, provide a more statistically conclusive evaluation.

Processing of soils and sediments at the Waukegan Harbor Site continued without further incident and was successfully completed on June 23, 1992 when a total of 12,700 tons had been processed. A summary of production data is provided in Table 3 (o page 3.9). Approximately 30,000 gallons of PCB con-

Table 2

EPA Superfund Innovative Technology Evaluation Draft Results

Waste Feed
Contaminant Concentrations

Contaminant	Average Concentration
Total PCBs	9231 ppm
Dioxin/Furan:	
TCDF	86 ppb
PeCDF	16 ppb

Treated Soils
Contaminant Concentrations

Contaminant	Average Concentration
Total PCBs	1.972 ppm
Dioxin/Furan:	
TCDF	5.4 ppb

Waste Discharge Oil
Contaminant Concentrations

Contaminant	Average Concentration
Total PCBs	32 %
Dioxin/Furan:	
TCDF	136 ppb
PeCDF	14 ppb

Stack Gas
Contaminant Concentrations

Contaminant	Average Concentration	Output (lbs/hr)	Emissions Criteria
Particulates	0.0039 ug/dscm	0.071	8.7 lb/hr
Total PCBs	0.8330 ug/dscm	1.7 E-5	
Total TCDF	0.0790 ng/dscm	1.4E-9	30 ng/m^3
HCL Gas	23.000 ug/dscm	0.00042	0.2 lb/hr
Total Hydrocarbons	ND	0	

Stack Gas Emissions
Destruction and Removal Efficiencies

Test Run No.	PCBs Fed to ATP (lb/hr)	PCBs Exiting Stack (lb/hr)	DRE[1] (%)
1	140.60	18.9E-6	99.999987
2	136.87	14.27E-6	99.999990
3	153.73	17.15E-6	99.999989
4	181.71	8.09E-6	99.999996

Note:
1. Project emissions criteria for PCBs is 99.9999% DRE.

Table 3

Production Data
Waukegan Harbor

Week Ending	Productions (Tons)	Average PCB in Feed Soil (ppm)	Average PCB Removal Efficiency [1] (%)
1-25-92	561	15,500	99.97
2-01-92	1,476	9,243	99.95
2-08-92	1,202	11,657	99.98
2-15-92	1,205	13,143	99.98
2-22-92	1,493	9,571	99.98
2-29-92	606	7,025	99.98
3-07-92	841	7,060	99.99
3-21-92	245	9,950	99.97
4-11-92	483	8,350	99.97
4-18-92	592	13,740	99.99
5-16-92	377	8,800	99.99
5-30-92	46	12,000	99.99
6-06-92	1,284	10,486	99.98
6-13-92	1,000	9,450	99.98
6-20-92	887	9,917	99.99
6-27-92	402	9,300	99.95
TOTAL	12,700	10,400	99.98

Note:
1. Soil treatment criterion is 97 percent removal efficiency.

taminated oil was desorbed from the soils and sediments and returned to the PRPs for subsequent off-site disposal. An average plant throughput of 8 tph had been achieved with a mechanical availability of 85 percent.

Process water generated during operations was discharged to Canonie's water treatment system at approximately 5 gallons-per-minute (gpm) and released to the sewer at a PCB concentration of less than 1 ppb PCBs.

D

APPENDIX D

List Of Vendors and Consultants

Chemical Waste Management
1950 South Batavia Ave.
Geneva, IL 60134-9838
(708) 513-4578

Progressive Recovery, Inc.
700 Industrial Dr.
Dupo, IL 62239
(618) 286-5000

Soil Purification, Inc
P.O. Box 72515
Chattanooga, TN 37407
(404) 861-0069

Texarome, Inc.
P. O. Box 157
Leakey, TX 78873
(210) 232-6079

Westinghouse Environmental & Geotechnical Services Inc.
111 Kelsey Lane, Suite B, #11
Edwinslack, FL 33619
(813) 620-1432

Roy F. Weston Inc.
1 Weston Way
West Chester, PA 19380-1499
(215) 430-7428 FAX (215) 430-3126

Site Reclamation System, Inc.
P.O. Box 11
Howey-In-The-Hills, FL 34737
(904) 324-3651

Nevada Hydrocarbon, Inc.
P.O. Box 9927
Reno, NV 89507
(702) 342-0200

Encore Environmental
344 West Henderson Road
Columbus, OH 43214
(614) 263-9287

Thermotech Systems Corporation
5201 North Orange Blossom Trail
Orlando, FL 32810
(407) 290-6000

Four Nines, Inc.
125 E. Trinity Pl., Suite 305
Decatur, GA 30030
(404) 370-0490

Tarmac Equipment
North 7 Highway
Blue Springs, MO 64014
(800) 833-4383

Focus Environmental, Inc.
9050 Executive Park Drive
Suite A-202
Knoxville, TN 37923
(615) 694-7517

Waste-Tech Services, Inc.
800 Jefferson County Parkway
Golden, CO 80401
(303) 279-9712

FB&D Technologies, Inc.
P.O. Box 58009
375 Chipeta Way
Salt Lake City, UT 84158-0009
(801) 583-3773

Ariel Industries
P.O. Box 9298
403 Spring Creek Road
Chattanooga, TN 37412
(615) 894-1957

International Technology (IT) Corp.
304 Directors Drive
Knoxville, TN 37923
(615) 690-3211

Canonie Environmental Services
800 Canonie Drive
Porter, IN 46304
(219) 926-8651

Separation and Recovery Sys., Inc.
1762 McGaw Ave.
Irvine, CA 92714-4962
(714) 261-8860

SoilTech, Inc.
94 Inverness Terrace East, Suite 100
Englewood, CO 80112
(303) 790-1410

ABB Environmental Services, Inc.
261 Commercial Street
P.O. Box 7050
Portland, ME 04112
(207) 775-5400

AAA Consulting Services Inc.
P.O. Box 5067
Novato, CA 94948
(415) 883-6380

Halliburton NUS
Environmental Corp.
5950 North Course Drive
P.O. Box 721110
Houston, TX 77272
(713) 561-1556

Soil Remediation Co.
P.O. Box 6217
Denver, CO 80206
(303) 756-2441
(800) 441-1968

Williams Environmental Services, Inc.
2076 West Park Place
Stone Mountain, GA 30087
(404) 498-2020

U.S. Waste Thermal Processing
3419 DiaLido, Suite 308
Newport Beach, CA 92663
(714) 509-7783

GDC Engineering
822 Neosho Avenue
Baton Rouge, LA 70802

Remedquip International Manufacturing
102B-267 West Esplanade
North Vancouver, BC V7M1A5

Remediation Technologies, Inc.
9 Pond Lane
Concord, MA 01742
(508) 371-1422

Southdown Thermal Dynamics
12235 FM 529
Houston, TX 77041
(800) 364-2402

E

APPENDIX E

Table Of Acronyms and Abbreviations

APC	Air Pollution Control
ARAR	Applicable or Relevant and Appropriate Regulations
ATP	Anaerobic Thermal Processor
BDAT	Best Demonstrated Available Technology
BIF	Boiler and Industrial Furnace
CERCLA	Comprehensive Environmental Response, Compensation, and Liability Act
DBA	Deutsche Babcock Anlagen
FRP	Fiberglass Reinforced Plastic
HCl	Hydrochloric Acid
LEL	Lower Explosion Limit
LTTA	Low Temperature Thermal Aeration Process
MGP	Manufactured Gas Plant
NO_x	Oxides of Nitrogen
NPDES	National Pollutant Discharge Elimination System
OSHA	Occupational Safety and Health Administration
PAH	Polynuclear Aromatic Hydrocarbon
PCB	Polychlorinated Biphenyls
PCE	Tetrachloroethene
PIC	Product of Incomplete Combustion

PNA	Polynuclear Aromatic Hydrocarbon
POTW	Publicly Owned Treatment Works
QA/QC	Quality Assurance / Quality Control
RCRA	Resource Conservation and Recovery Act
RI/FS	Remedial Investigation / Feasibility Study
ROD	Record of Decision
SARA	Superfund Amendments and Reauthorization Act
SCR	Silicon Controlled Rectifier
SITE	Superfund Innovative Technology Evaluation Program
SO_x	Oxides of Sulfur
SO_2	Sulfur Dioxide
SVOC	Semi-Volatile Organic Compound
TCA	1,1,1 Trichloroethane
TCDD	Tetrachlorodibenzo-p-dioxin
TCDF	Tetrachlorodibenzofuran
TCE	Trichloroethene
TCLP	Toxicity Characteristic Leachate Procedure
TSCA	Toxic Substance Control Act
USA	United States of America
US EPA	United States Environmental Protection Agency
UST	Underground Storage Tank
VOC	Volatile Organic Compound

F

APPENDIX F

List Of References

Alperin, E.S., A. Groen, and R.W. Helsel. 1992. Thermal Desorption of PAH-contaminated soils. In *Proc. 1992 Industrial Pollution Control Conference.* Atlanta. Feb. 9-12.

Ayen, R.J. and C.P. Swanstrom. 1992. Low temperature thermal treatment for petroleum refinery waste sludges. *Environmental Progress* 11:127.

Barton, R.G., W.D. Clark, and W.R. Seeker. 1990. Fate of metals in waste combustion systems. *Combust. Sci. Technol.* 74:327.

Bell, B.B. and R.G. Giese. 1991. Effective low-temperature thermal aeration of soils. *Remediation* 1:461-72.

Borkent-Verhage, C., D. Cheng, L. deGalan, and Ed W.B. deLeer. 1986. Thermal cleaning of soil contaminated with g-hexachlorocyclohexane. In *Contaminated Soils*, ed. J.W. Assink and W.J. van den Brink. Dordrecht, Netherlands: Martinus Nijhoff Publishers.

Bruner, C.R. 1985. *Hazardous air emissions from incineration.* New York: Chapman and Hall.

California Department of Health Services, Toxic Substances Control Program, Alternative Technology Division. 1990a. *Remedial technology demonstration project report: soil cleanup system for a diesel contaminated site in Kingvale, California.* Staff Report. Jan.

California Department of Health Services, Toxic Substances Control Program, Alternative Technology Division. 1990b. *Thermal treatment process for fuel contaminated soil: U.S. waste thermal processing.* Staff Report. March.

Canonie Environmental Service Corp. 1991. *Low temperature thermal aeration soil remediation services.* Project 77-305. Porter, Ind.

Cudahy, J.J. and W.L. Troxler. 1992. 1991 thermal remediation industry survey. *J. Air & Waste Manage. Assoc.* 42:844.

Flytzani-Stephanopoulos, M., A.F. Sarofim, L. Tognotti, H. Kopsinis, and M. Stoukides. 1991. Incineration of contaminated soils in an electrodynamic balance. In *Emerging technologies in hazardous waste management II*. American Chemical Society.

Fox, R.D., E.S. Alperin, and H.H. Huls. 1991. Thermal treatment for the removal of PCBs and other organics from soil. *Environmental Progress* 10:1.

Gilot, P., M. Koch, H. Saito, V. Bucala, J.B. Howard, and W.A. Peter. 1992. Fundamental studies of thermal methods for the decontamination of soil. Poster at the *Science Advisory Committee Meeting of the Northeast Hazardous Substance Research Center*. New Jersey Institute of Technology. April.

Gorte. R.J. 1982. Design parameters for temperature programmed desorption from porous catalysts. *Journal of Catalysis* 75:164.

Halliburton NUS Environmental Corporation. 1991. *Ecotechniek thermal treatment system* (Marketing Literature). Houston.

Harwood, E. 1992. Comparing thermal counter flow and parallel flow soil remediation processes. In *Proc. 11th International Incineration Conference.* Albuquerque, New Mexico. May 11-15.

Helsel, R.W., E. Alperin, and A. Groen. 1989. *Engineering-scale demonstration of thermal desorption technology for manufactured gas plant site soils.* HWRIC RR-038. Hazardous Waste Research and Information Center. Savoy, Ill.

Helsel, R.W. and A. Groen. 1988. *Laboratory study of thermal desorption treatment of contaminated soils from former manufactured gas plant sites.* GRI - 88/0161. Chicago. Gas Research Institute. August.

Herz, R.K., J.B. Kiela, and S.P. Marin. 1982. Adsorption effects during temperature-programmed desorption of carbon monoxide from supported platinum. *Journal of Catalysis* 73:66.

Hutton, J.H. and R. Shanks 1993. Thermal desorption of PCB contaminated waste at the Waukegan Harbor superfund site-a case study. Presented at *Innovative Treatment Technologies-Uses and Applications for Site Remediation, Thermal I-Thermally Enhanced Volatilization,* Satellite Seminar. Feb. 18.

Jones, D.M. and G.L. Griffin. 1983. Saturation effects in temperature-programmed desorption spectra obtained from porous catalysts. *Journal of Catalysis* 80:40.

Keyes, B.R. 1992. A fundamental study of the thermal desorption of organic wastes from montmorillonite clay particles. Ph.D. diss., University of Utah.

Kulwiec, R.A., ed. 1985. *Materials handling handbook*, 1032. 2d ed. John Wiley and Sons.

Lauch, R.P., J.G. Herrmann, M.L. Smith, E. Alperin, A. Groen, and J. Hessling. 1991. Removal of creosote from soil by thermal desorption. In *Proc. Hazard-*

ous Materials Control Research Institute's 12th Annual National Conference and Exhibition. Washington, D.C. Dec. 3-5.

Lehmann, B.G. 1991. Experience with the decontamination of the Konigsborn coke-oven plant site using a pyrolysis plant. In *Proc. Fifth International Conference, North Atlantic Treaty Organization, Committee on the Challenges of Modern Society (NATO/CCMS)*. Washington, D.C. Nov. 18-22.

Lester, T.W., V.A. Cundy, A.M. Sterling, A.N. Montestruc, A.L. Jakway, C. Lu, C.B. Leger, D.W. Pershing, J.S. Lighty, G.D. Silcox, and W.D. Owens. 1991. Rotary kiln incineration: comparison and scaling of field-scale and pilot-scale contaminant evolution rates from sorbent beds. *Environ. Sci. Technol.* 25:1142.

Lide, D.R., ed. 1990. *CRC handbook of chemistry and physics*. 71st ed. Boston: CRC Press Inc.

Lighty, J.S., D.W. Pershing, V.A. Cundy, and D.G. Linz. 1988. Characterization of thermal desorption phenomena for the cleanup of contaminated soil. *Nuclear and Chemical Waste Management* 8:225.

Lighty, J.S., G.D. Silcox, D.W. Pershing, V.A. Cundy, and D.G. Linz. 1990. Fundamentals for the thermal remediation of contaminated soils: particle and bed desorption models. *Environ. Sci. Technol.* 24:750.

Lighty, J.S., E.G. Eddings, E.R. Lingren, X.X. Deng, D.W. Pershing, R.M Winter, and W.H. McClennen. 1990. Rate limiting processes in the rotary kiln incineration of contaminated solids. *Combust. Sci. and Tech.* 74:31.

Lighty, J.S., G.D. Silcox, and D.W. Pershing. 1990. *Investigation of rate processes in the thermal treatment of contaminated soils*. GRI-90/0112.Gas Research Institute. Chicago. March.

Locke, B.B., M.M. Arozarena, and C.D. Chambers. 1991. *Evaluation of alternative treatment technologies for CERCLA soils and debris (Summary of Phase I and Phase II)*. EPA/600/14. US EPA: Cincinnati.

McCormick, R.J., R.J. Derosier, K. Lim, R. Larkin, and H. Lips. 1985. *Costs for hazardous waste incineration: capital operation and maintenance, retrofit*. Park Ridge, N.J.: Noyes Publications.

Owens, W.D., G.D. Silcox, J.S. Lighty, X.X. Deng, D.W. Pershing, V.A. Cundy, C.B. Leger, and A.L. Jakway. 1991. Thermal analysis of rotary kiln incineration: comparison of theory and experiment. *Combust. and Flame* 86:101.

PRC Environmental Management, Inc., Versar, Inc., and Radian Corporation. 1991. *Canonie Environmental Services Corporation low temperature thermal aeration (LTTA) process site demonstration at the Anderson Development Company site Adrian, Michigan: draft demonstration plan, quality assurance project plan*. Prepared for US EPA. June.

Rogers, J.A., T.M. Holsen, and P.R. Anderson. 1990. Effect of humidity on low temperature thermal desorption of VOCs from contaminated soil. In *Proc. ASCE National Conference on Environmental Engineering*. Arlington, Va.

Romzick, P.G. and C. Swanstrom. 1991. *X*TRAX™ Laboratory treatability study of jet fuel contaminated soil from Chanute Air Force Base near Rantoul, Illinois*. HWRIC TR-003. Hazardous Waste Research and Information Center. Champaign, Ill. November.

Roenzweig, M.D., ed. 1991. Dallas meeting highlights environmental advances. *Chemical Engineering Progress* August: 9.

Sax, N. 1989. *Dangerous properties of industrial materials*. 7th ed. New York: Van Nostrand Reinhold.

Schneider, D. and B.D. Beckstrom. 1990. Cleanup of contaminated soils by pyrolysis in an indirectly heated rotary kiln. *Environmental Progress* 9(3): 165.

Scotto, M.V., T.W. Peterson, and J.O.L. Wendt. 1992. Hazardous waste incineration: the in-situ capture of lead by sorbents in a laboratory downflow combuster. In *Proc. Twenty-fourth (International) Symposium on Combustion*, 867. The Combustion Institute. Pittsburgh.

Silcox, G.D. and D.W. Pershing. 1990. The effects of rotary kiln operating conditions and design on burden heating rates as determined by a mathematical model of rotary kiln heat transfer. *J. Air & Waste Manage. Assoc.* 40:337.

Soil Purification Inc. 1991. ASTEC marketing literature. WMS 2M. Nov.

Swanstrom, C. 1991. Determining the applicability of X*TRAX™ for on-site remediation of soil contaminated with organic compounds. In *Proc. Hazmat '91 Central*. Rosemont, Ill. April 3.

Swanstrom, C. and C. Palmer. 1990. X*TRAX - Transportable thermal separator for organic contaminated solids. In *Proc. Second Forum on Innovative Hazardous Waste Treatment Technologies: Domestic and International*. US EPA. Philadelphia. May 15-17.

Swanstrom, C. and C. Palmer. 1991. Thermal separation of solids contaminated with organics. In *Proc. Hazmat '91 West*. Long Beach, Calif. Nov. 19-21.

Tarmac Equipment Company. *Review of issues concerning soil recycling facilities*. (Marketing Literature). Kansas City.

Troxler, W.L. , J.J. Cudahy, R.P. Zink, S.I. Rosenthal, and J.J. Yezzi. 1992. Treatment of petroleum contaminated soils by thermal desorption technologies. Presented at *85th Annual Meeting of 1992 Air and Waste Management Association*. Kansas City. June 21-26.

Turner, C.F. and J.W. McCreery. 1981. *The chemistry of fire and hazardous material*. Boston: Allyn and Bacon.

Uberoi, M. and F. Shadman. 1990. Sorbents for the removal of lead compounds from hot flue gases. *AIChE Journal* 36(2): 307-309.

Uberoi, M. and F. Shadman 1991. High temperature removal of cadmium compounds using sorbents. *Environ. Sci. Technol.* 25(7): 1285-9.

US EPA. 1987. *Superfund record of decision, Ottati & Goss/Great Lakes, NH, first remedial action-find.* EPA/ROD/ROI-87/021. Washington, D.C.

US EPA. 1991a. *The superfund innovative technology evaluation program: technology profiles.* 4th ed. EPA/540/5-91/008. Washington, D.C. Nov.

US EPA. 1991b. *Engineering bulletin - thermal desorption treatment.* EPA/540/2-91/008. Cincinnati, Ohio. May.

US EPA. 1991c. *Superfund site interim close out report, Cannons Engineering Corporation (CEC) site, Bridgewater, Mass.* Sept. 30.

US EPA. 1992a. *Air emissions from the treatment of soils contaminated with petroleum fuels and other substances.* EPA-600/R-92-124, OAR and ORD. Washington D.C. July.

US EPA. 1992b. *Alternative Treatment Technology Information Center (ATTIC)* (data base). Office of Solid Waste and Emergency Response, Technology Innovation Office. Washington, D.C.

US EPA. 1992c. *Application Analysis Report WESTON Low Temperature Thermal Treatment Technology.* In Review.

US EPA. 1992d. *Guide to conducting treatability studies under CERCLA: thermal desorption remedy selection — interim guidance.* 540/R-92/074A. RREL Cincinnati, and OERR and OSWER, Washington, D.C.

US EPA. 1992e. *Innovative treatment technologies: semi-annual status report.* 3rd ed. EPA/540/2-91/001. Office of Solid Waste and Emergency Response. Washington, D.C. April.

US EPA. 1992f. *Potential reuse of petroleum contaminated soil-a directory of permitted recycling facilities.* EPA/600/R-92/096. June.

US EPA. 1992g. *Thermal desorption applications manual for treating petroleum contaminated soils.* EPA Report. Contract No. 68-C9-0033. RREL. Edison, N.J. In Review. (At this writing, this Guidance Document is a draft that has not been formally released by the EPA. It should not at this stage be construed to represent agency policy.) Jan.

US EPA. 1992h. *Vendor Information System for Innovative Treatment Technologies (VISITT)* Version 1.0 (data base). EPA/542/R-92/001 No. 1. Office of Solid Waste and Emergency Response, Technology Innovation Office. Washington, D.C. June.

US EPA. 1993. *Engineering bulletin-thermal desorption on treatment.* Vol. 2. EPA/540/0-00/000. OERR, Washington D.C., and ORD, Cincinnati. In review.

Varuntanya, C.P., M. Hornsby, A. Chemburkar, and J.W. Bozzelli. 1989. Thermal desorption of hazardous and toxic organic compounds from soil matrices. In *Petroleum Contaminated Soils*, 251. Chelsea, Mich.: Lewis Publishers.

Vorum, M. 1991. Dechlorination of polychlorinated biphenyls using the SoilTech anaerobic thermal processing unit; Wide Beach superfund site, New York. In *Proc. HazTech International 91*. Pittsburgh. May 14-16.

Vorum, M. 1992. SoilTech anaerobic thermal process (ATP): rigorous and cost-effective remediation of organic-contaminated solid and sludge wastes. Paper 92-31.07 presented at *85th Annual Meeting of Air and Waste Management Association*. Kansas City. June 21-26.

Vorum, M. and R.M. Ritcey. 1991. SoilTech ATP system: pyrolytic decontamination of oily RCRA and TSCA wastes. In *Proc. HazTech International 91*. Houston. Feb. 13.

Wark, K. and C.F. Warner. 1981. *Air pollution: its origin and control*. 2d. ed. New York: Harper Collins Publishers.

Wu, Y.G. and J.W. Bozzelli. 1992. Mass transfer studies related to thermal adsorption-desorption of C_6H_6 and C_6H_5Cl on soil matrices. Submitted to *Environmental Science and Tech.*

Wu. Y.G., J. Dong, and J.W. Bozzelli. 1992. Mass transfer of hazardous organic compounds in soil matrices: experiment and model. *Combust. Sci. and Tech.* 83:151-63